Polymer and Food Rheology

The science of rheology and its applications have grown steadily. Although polymers dominated in the past, the applications to foods have become increasingly important over the last four decades. The complexity of food rheology has not only generated increased interest among academics, but it has also led to an increased recognition of its importance in the food industry. Emphasizing physics rather than mathematics, *Polymer and Food Rheology: Volume 1, Applications to Single-Phase Systems* focuses on rheological characteristics and properties of single substances and homogeneous mixtures. It introduces fundamental principles, types of flows, types of instruments, their theory, capabilities, limitations, and sources of error. Both sophisticated rheometers and simple devices and their uses are considered.

- This book discusses the types of material functions (properties) obtained from the different types of measurements and techniques: shear viscosity, normal stress differences, dynamic mechanical properties, and extensional viscosity.
- It uses selected data of polymers and foods to illustrate the types of behaviors one could expect from these materials.
- It shows the effect of polymer structure, such as molecular weight, molecular weight distribution, long chain branching, temperature, pressure, and deformation rate on the material properties, using published data on LDPE, HDPE, and LLDPE; the applicability of these relationships to foods is also discussed.
- It links the measurement techniques to practical applications such as the choice, processing, and design of polymers and foods.
- It explains concepts such as entanglement, shear thinning, extensional strain hardening, melt strength, time-temperature superposition, among others and their relevance to practical applications.
- It describes the theory behind selected constitutive equations, their utility in modeling, and applications.

This text is aimed at researchers and industrial practitioners in materials and chemical engineering as well as chemistry.

Polymer and Food Rheology

Volume 1, Applications to Single-Phase Systems

Third Edition

by Rakesh K. Gupta and Mahesh Padmanabhan

CRC Press
Taylor & Francis Group
Boca Raton London New York

CRC Press is an imprint of the
Taylor & Francis Group, an **informa** business

Designed cover image: © 2026 Shutterstock

Third edition published 2026
by CRC Press
2385 NW Executive Center Drive, Suite 320, Boca Raton FL 33431

and by CRC Press
4 Park Square, Milton Park, Abingdon, Oxon, OX14 4RN

CRC Press is an imprint of Taylor & Francis Group, LLC

First edition published by Marcel Dekker 1977
Second edition published by Marcel Dekker 2000

Library of Congress Cataloging-in-Publication Data
Names: Gupta, Rakesh K., (Rakesh Kumar), 1954- author. | Padmanabhan, Mahesh, author.
Title: Rheology of complex fluids and soft matter : with applications to
polymers and foods / Rakesh K. Gupta, Mahesh Padmanabhan.
Other titles: Polymer and composite rheology
Description: Third edition. | Boca Raton : CRC Press, 2026- | The first edition in 1977 was
entitled Polymer rheology written by Lawrence E. Nielsen; the second edition was written by
Rakesh K. Gupta entitled Polymer and composite rheology; this new edition in two volumes
with co-author Mahesh Padmanabhan. | Includes bibliographical references and index.
Identifiers: LCCN 2025040455 (print) | LCCN 2025040456 (ebook) | ISBN 9780367438470
(v. 1 ; hardback) | ISBN 9781041224419 (v. 1 ; paperback) | ISBN 9781003013501 (v. 1 ; ebook)
Subjects: LCSH: Polymers--Rheology. | Complex fluids--Fluid dynamics. |
Soft condensed matter--Fluid dynamics. | Food--Analysis. | Polymeric composites--Fluid dynamics.
Classification: LCC TA455.P58 G88 2026 (print) | LCC TA455.P58 (ebook)
LC record available at https://lccn.loc.gov/2025040455
LC ebook record available at https://lccn.loc.gov/2025040456

ISBN: 978-0-367-43847-0 (hbk)
ISBN: 978-1-041-22441-9 (pbk)
ISBN: 978-1-003-01350-1 (ebk)

DOI: 10.1201/9781003013501

Typeset in Times
by KnowledgeWorks Global Ltd.

To my daughters Deepti and Neha and my grandsons Shreyas and Aadav, whose curiosity and joy continue to inspire my own learning.

—Rakesh K. Gupta

To my wife, Deepa, my sister, Geetha, and
in loving memory of my parents.

—Mahesh Padmanabhan

May we walk together
May we speak in unison
May we strive together
May our hearts be in common
Rig Veda 10/191/1-4 (Hindu scripture)

Contents

Authors

Rakesh K. Gupta is Professor and George B. & Carolyn A. Berry chair in the Department of Chemical and Biomedical Engineering at West Virginia University (WVU), where he has been teaching since 1992. He received his Ph.D. degree from the University of Delaware in 1980 and his B.Tech. degree from the Indian Institute of Technology, Kanpur, India, in 1975, both in chemical engineering. Before coming to WVU, he taught at the State University of New York at Buffalo for almost 11 years. He has also worked briefly for Monsanto, DuPont, and Braskem America. His research is focused on the processing and rheology of polymers and foods and also on polymer composites and nanocomposites. He has authored or coauthored over 135 journal papers, 80 conference proceedings papers, and 22 book chapters on these topics. He also holds four U.S. patents. He is the coauthor of the textbook *Fundamentals of Polymer Engineering* (now in the third edition) and coeditor of (1) *Handbook of Polymer Nanocomposites* and (2) *Graphite, Graphene and Their Polymer Nanocomposites*. While his polymer rheology research has been supported by a variety of federal, state, and private sources, his food rheology research has primarily been funded by Kraft Foods and Mondelēz International. He is a member of the Editorial Board of the *Journal of Vinyl and Additive Technology*, and he is a Fellow of the Society of Plastics Engineers.

Mahesh Padmanabhan is a Research Fellow in the areas of Rheology, Texture, and Physical Sciences at Mondelēz International, based in East Hanover, New Jersey; he also holds an Adjunct Professor of Chemical Engineering appointment at West Virginia University, Morgantown, West Virginia. He has been with Mondelēz (previously Kraft Foods and Nabisco Foods) for over 28 years, applying the principles of rheology and physical sciences, developing methods and insights for a range of industrial problems and food systems: from ingredients, formulation, structure-property relations, processing, manufacturing, quality, to product texture related to consumer perception. Before joining Mondelēz, Dr. Padmanabhan led the Rheology Lab of the Adhesives Division at the National Starch & Chemical Company, Bridgewater, New Jersey, following a postdoctoral fellowship in chemical engineering at the University of Minnesota. He received his B.Tech. (Hons.) degree (1985) from the Indian Institute of Technology, Kharagpur, India, M.S. (1989) and Ph.D. (1992) degrees from the University of Minnesota (St. Paul), all in agricultural engineering, and an M.S. (1993) in chemical engineering, also from the University of Minnesota (Minneapolis).

Preface

The first edition of the book appeared in 1977, and it was authored by the late Larry Nielsen under the title *Polymer Rheology*. Nielsen had a Ph.D. in physical chemistry, and his professional career, spanning 32 years, was spent working for the Monsanto Company from which he retired as Distinguished Science Fellow. As explained by him in the preface to the first edition, there was, at that time, no book on the subject that was suitable for the majority of people who needed to know something about rheology to properly do their job. The book served that purpose and stood the test of time, and the second edition was written by Rakesh K. Gupta in 2000 as *Polymer and Composite Rheology*. The guiding principle remained the same, but the scope was expanded to provide equal coverage to single-phase and multiphase systems; the latter included suspensions, emulsions, foams, and powders.

Much has changed in the last 25 years or so. These changes include the synthesis of polymers whose molecular weight can be precisely controlled, advancement in rheometers with the addition of more measurement (such as powder characterization), and advanced data acquisition capabilities allowing for the performance and analysis of tests such as large amplitude oscillatory shear (LAOS) flow, and continued development of tube models of polymer behavior. Thus, much new information, both practical and theoretical, has become available on the flow behavior of a variety of materials of interest. This, by itself, merits the publication of a new edition of the book. In addition, the field of polymer rheology has changed. Fifty years ago, one could attend a rheology meeting and expect to understand most of what was being presented. Such is not the case today. Lectures and journal articles have become very specialized and highly mathematical. Simultaneously, the lecturers, for the most part, come from academia, but the audience still has a significant industrial component; applications of rheology remain a critical area for many industries. The purpose of the new edition is also to cater to those who are not specialists in polymer rheology but who require an understanding of the subject in order to be able to do their job.

Over the years, foods, pharmaceuticals, and consumer products have gained prominence in the rheology community. Foods, after all, are fats, proteins, and carbohydrates, and these are all polymers of varying molecular weights, but the products are much more complex. Food processing is similar to polymer processing, and similar equipment is used. As we show in this edition of the book, there are immense similarities between polymers and foods, and the two can be studied using the same framework. A difference between the two areas, though, is that there is immense industrial interest and industrial participation in research on food rheology. This is very evident to those who attend the International Congress on Rheology. Another major difference is that foods are eventually consumed, which introduces a new set of problems for the rheologist to measure and quantify. Given the world-wide trend toward healthy foods, we anticipate that food rheology will be a source of challenging and meaningful problems in the future for rheologists to work on. In view of this, we have chosen to devote as much effort to food rheology as polymer rheology.

We have also chosen to divide the book into two volumes. The first volume is on single-phase systems, while the second volume is on two-phase systems. Since most food items are composed of two or more phases, the second volume allows us to do justice to food rheology. Mahesh Padmanabhan of Mondelēz International is an expert on food rheology, and he is now a coauthor of the third edition.

The specific aims of the third edition are the same as those of the previous editions, and the format is similar. We talk about why we make rheological measurements, how we do this in different well-defined flow fields, what do data look like, how do we interpret these data in light of current theories, how do we represent them with simple equations, and how do we relate one measurement to another. Of course, we continually emphasize the utility of rheological theories and measurements. This book is based on the experiences of three different researchers, and, given that Nielsen earned his Ph.D. in 1945, this book covers developments essentially from the time of inception of the field of rheology. This book also has a very industrial orientation since Gupta has worked in industry and has also had unbroken industrial funding since he became an academic. Padmanabhan's entire professional career has been in industry, and he now works for a major food company.

We have been passionate about the topics that we have written about, and we hope that this is evident to the readers. We also hope that we have succeeded in our goal of making rheology accessible to those who are not already initiated in this subject.

Rakesh K. Gupta

Mahesh Padmanabhan

1 Introduction to Rheology

I. RHEOLOGY

Rheology can be defined as the science of the flow and deformation of matter. For low-molecular-weight fluids, the study of rheology involves the measurement of viscosity. For such fluids, the viscosity depends primarily upon the temperature and hydrostatic pressure. However, the rheology of complex fluids and soft matter (such as polymeric fluids, whether neat or filled, foods, cosmetics, and other industrial fluids) poses many challenges due to their non-ideal behavior. In addition to having complex shear viscosity behavior, these liquids show elastic properties, such as unequal normal stresses in shear and a prominent tensile viscosity in extension. All these rheological properties depend upon the rate of deformation, the molecular weight and structure of the polymer, and the concentration of various additives, fillers, and immiscible liquids, as well as upon the temperature. In addition, even at a constant rate of deformation, stresses are found to depend on time that may be either reversible or irreversible.

The subject of rheology is very important for both polymers and polymeric composites. This is true for two reasons. First, flow is involved in the processing and fabrication of such materials in order to make useful objects. Thus, fluid rheology is relevant to polymer processing and determines stress levels in operations such as extrusion, calendering, fiber spinning, and film blowing. Similarly, rheology influences residual stresses, cycle times, and void content in composite processing operations such as bag molding, compression molding, and injection molding. Clearly, a quantitative description of polymer and composite rheology is essential for developing models of the various polymer processing operations; these models can be employed for process optimization and for predicting the onset of flow instabilities. In the use of polymers, though, it is generally the mechanical properties that are important. Mechanical behavior, however, is influenced by rheological behavior, and this is the second important reason for studying polymer rheology. For example, molecular orientation has dramatic effects on the mechanical properties of molded objects, fibers, and films. For short-fiber composites, fiber orientation plays the role of molecular orientation in unfilled systems. The kind and degree of molecular or fiber orientation are largely determined by the rheological behavior of the polymer and the nature of the flow in the fabrication process.

Rheology is involved in many other aspects of polymer science. For example, many polymers are made from emulsions of monomers in stirred reaction vessels. The resulting latices flow through pipes and may end up as a paint that is applied to a surface by some process in which the rheological properties of the latex must be controlled carefully. Plastisols, which are a suspension of a polymer in a liquid, are fabricated into useful objects by processes such as rotational molding. Powdered polymers or granules must flow from bins and must perform properly in a fabrication

DOI: 10.1201/9781003013501-1

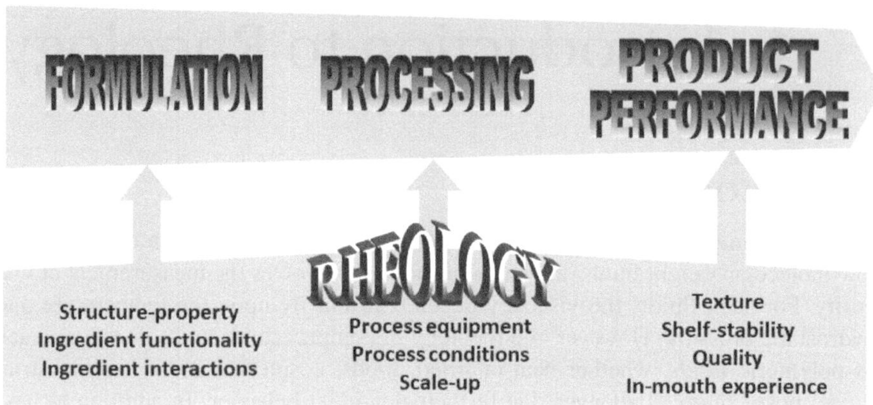

FIGURE 1.1 Illustration of the breadth of importance of rheology in the food industry. (M. Padmanabhan, unpublished figure.)

process such as rotational molding and in powder coatings. The rheology of polymer powders is important also in the first sections of extruders and in injection molding machines before the polymer softens to a liquid. Note that at a less fundamental level, rheological measurements, both transient and steady state, can be employed for product characterization and quality control purposes. Such measurements are often also used for examining and understanding the interaction of the different constituents of a multicomponent or multiphase mixture and their influence on the flow and other properties of such materials.

In general, foods are complex and usually not thermodynamically in equilibrium. The large variability (with geography and time) of nature-sourced ingredients used in food manufacture exaggerates this complexity. Foods are commonly a complex mixture of natural polymers, emulsions, suspensions, and entrapped air.

Rheology and its principles are of great importance to foods in several ways (Figure 1.1): understanding of ingredient properties governing their choice, the properties of formulations due to the ingredients and the interactions between them, the flow behavior and irreversible changes during processing, the mechanical properties of the food products for packaging, quality and shelf-stability, and finally and perhaps most significantly the in-mouth breakdown during mastication delivering desirable and satisfying eating experience to consumers.

II. MATERIAL FUNCTIONS IN VISCOMETRIC FLOWS

The flow field that is generated in most standard instruments used to measure rheological properties is a particular kind of shear flow called *viscometric flow*. All the motion in a viscometric flow is along one coordinate direction, say, x_1 in Figure 1.2, the velocity varies along a second coordinate direction, say, x_2, and the third direction is neutral. This is illustrated schematically in Figure 1.2, where a liquid is confined between two flat plates of area A separated by a distance D. When the upper

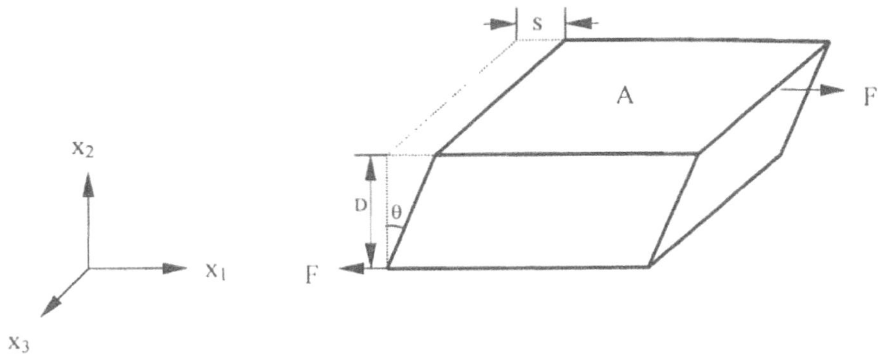

FIGURE 1.2 Schematic diagram for the measurement of shear viscosity. (From Ref. [1].)

plate is moved in the x_1 direction relative to the lower plate, the liquid is sheared with the amount of *shear strain* γ being defined as

$$\gamma = \tan\theta = \frac{\text{Amount of shear displacement } S}{\text{Distance between shearing surfaces } D} \tag{1.1}$$

The rate of shear strain will then be

$$\dot{\gamma} = \frac{\text{Relative velocity}}{\text{Distance } D} \tag{1.2}$$

which equals dv_1/dx_2 for small values of D. The rate of shear strain is commonly referred to as the *velocity gradient* or more simply as just the *shear rate.* It is evident that the shear rate will be independent of time if the relative velocity is constant.

A constant force F is typically required to move the top plate in Figure 1.2 at a constant velocity relative to the lower plate. This is a shear force, and on dividing this quantity by the area A of the shear face, one gets the *shear stress* τ. For a Newtonian liquid, the shear stress is directly proportional to the shear rate, with the constant of proportionality being called the *dynamic viscosity* η. If the fluid is not Newtonian, a plot of the shear stress against the shear rate is not a straight line but a curve, such as the lower solid line shown in Figure 1.3. The liquid may be Newtonian at very low shear rates to give a limiting viscosity η_0 from the initial slope of the τ-versus-$\dot{\gamma}$ curve. When the τ-versus-$\dot{\gamma}$ curve is nonlinear, the viscosity (the adjective *dynamic* is commonly omitted) is no longer a constant, and it becomes a function of the shear rate. It is now defined as the slope of the secant line from the origin to the shear stress at the given value of the shear rate; that is,

$$\eta = \frac{\tau}{\dot{\gamma}} \tag{1.3}$$

Another kind of non-Newtonian behavior is *Bingham plastic* behavior. This is illustrated by the upper solid line in Figure 1.3. Here the τ-versus-$\dot{\gamma}$ curve is linear,

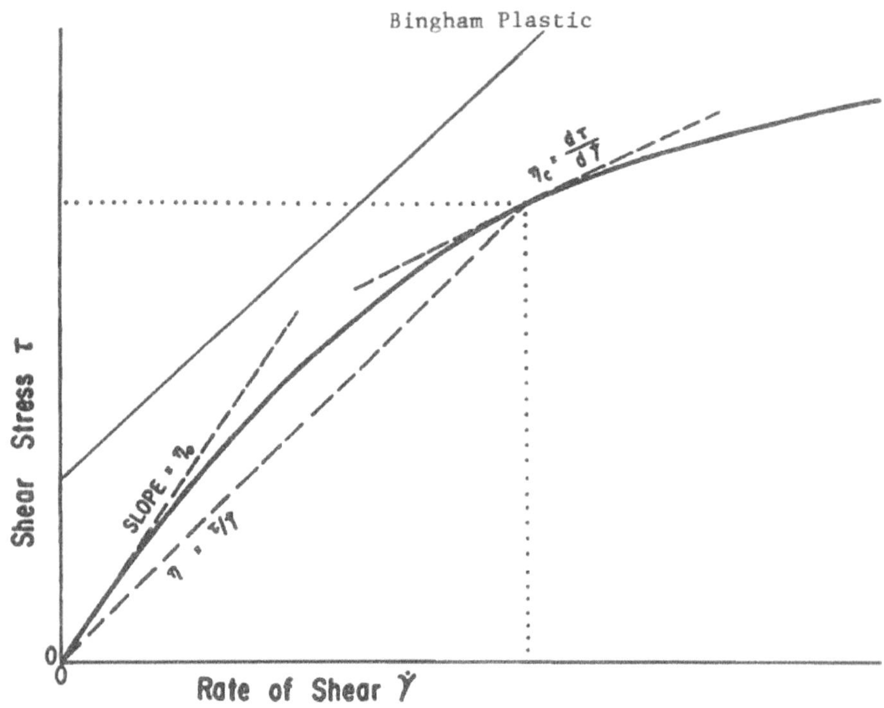

FIGURE 1.3 Measurement of viscosity from shear stress versus rate of shear curves. (From Ref. [1].)

but it does not pass through the origin; the value of the intercept on the shear stress axis is called the *yield stress*. Note that the slope of the shear stress curve at the chosen value of the shear rate is known as the consistency η_c.

The SI unit for viscosity is Pa-s, where Pa is the abbreviation for Pascal, which is Newtons per square meter. In the past, it was common to measure viscosity in the CGS unit of poise, which is the same as dyne second per cm^2 or g/cm-s. The viscosity of water is about 1 m Pa-s, or 1 centipoise, since 1 Pa-s equals 10 poise. Typical polymer melts have viscosities generally of the order of 10^2–10^3 Pa-s.

In general, it is found that the shear viscosity of polymers decreases as the shear rate increases. This behavior is known as *shear thinning*, and an important consequence of this result is that the well-known Hagen–Poiseuille equation does not hold for the flow of polymers through a circular tube. In particular, one finds that on doubling the flow rate through the tube, the necessary pressure drop is less than doubled.

For a fluid at rest, it is easy to show [2] that all the surface stresses that are present within the fluid are normal stresses and that at any location, these are not only equal in magnitude but also independent of the orientation of the surface. This single normal stress is compressive in character, and it is commonly called the *pressure* and denoted by the symbol p. From the definition of a fluid, the absence of motion of course implies the absence of shear stresses. Conversely, for a fluid in

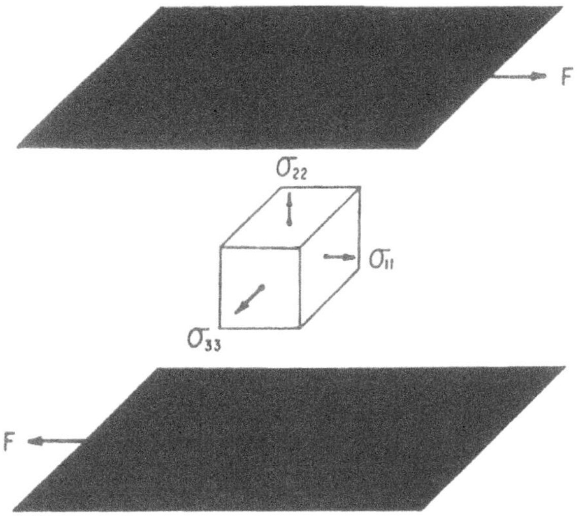

FIGURE 1.4 Schematic diagram showing the notation for normal stresses in a shear field. (From Ref. [1].)

motion, it must necessarily be true that either shear stresses are present or the normal stresses are unequal. For the situation depicted in Figure 1.2, we have seen that flow arises due to the application of a shear stress. If the liquid being sheared is polymeric, one finds that the normal stresses along the three coordinate directions (see Figure 1.4) become unequal; for Newtonian fluids, however, normal stresses remain equal to each other. Due to the fact that most liquids can be considered incompressible for all practical purposes, the application of equal normal stresses (or pressure) along the three coordinate directions does not lead to deformation or a change in volume; it is only unequal normal stresses that cause motion and deformation. As a consequence, one cannot determine the absolute value of the pressure from a measurement of fluid deformation, which is a measure of the change in distances between material points or material planes; one can only deduce normal stress differences.

It is common practice to define the first and second normal stress differences as follows:

$$\textit{First normal stress difference } N_1 = \sigma_{11} - \sigma_{22} \tag{1.4}$$

$$\textit{Second normal stress difference } N_2 = \sigma_{22} - \sigma_{33} \tag{1.5}$$

While the first normal stress difference tends to force the shear plates in Figure 1.4 apart, the second normal stress difference tends to create bulges in the polymer at the edge of the plates either parallel or perpendicular to the direction of the applied force *F*. Indeed, it is the presence of the first normal stress difference in shear that is responsible for the swelling of an isothermal jet of polymer on emerging into air from a die or capillary. It is found that the normal stress differences are even functions of the shear rate, and it is customary to define the first and second *normal stress coefficients*, Ψ_1, and Ψ_2, as ratios of the respective stress differences with the square of the shear rate. At low shear rates, both these coefficients tend to attain constant values.

In cases where the relative velocity of the shearing plates in Figure 1.4 is not constant but varies in a sinusoidal manner so that the shear strain and the rate of shear strain are both cyclic and have a small amplitude, the shear stress is also sinusoidal. In these dynamic mechanical experiments, the stress is out of phase with the rate of strain unless the fluid is Newtonian. In this situation, a *complex viscosity* η^* can be measured. In general, the complex viscosity contains an elastic component in addition to a term similar to the ordinary steady-state viscosity. The complex viscosity is defined by:

$$\eta^* = \eta' - i\eta'' \tag{1.6}$$

The *dynamic viscosity* η' is related to the steady-state viscosity and is the part of the complex viscosity that measures the rate of energy dissipation. Similarly, the imaginary viscosity η'' measures the elasticity, or stored energy. These two viscosities are computed from the real and imaginary parts of the shear modulus using the following relations:

$$\eta'' = \frac{G'}{\omega} \tag{1.7}$$

$$\eta' = \frac{G''}{\omega} \tag{1.8}$$

where ω is the frequency of the oscillations in radians per second. The *storage modulus*, or dynamic rigidity, G', is defined as the component of the stress in phase with the strain divided by the strain amplitude; the *loss modulus*, G'', is defined as the component of the stress out of phase with the strain divided by the strain amplitude. The ratio of the loss modulus to the storage modulus, G''/G', is known as tan δ and is an alternate measure of energy dissipation in viscoelastic materials. Note that for a perfectly elastic material, G'' is zero, while for Newtonian liquids (that are perfectly viscous) G' is zero. In general, both G' and G'' are functions of frequency.

If the amplitude of oscillation is large, the shear stress, while still cyclic, becomes non-sinusoidal and shows the presence of higher harmonics. This flow field is known as *large amplitude oscillatory shear* (LAOS). Now, even if the rheometer calculates G' and G'', these moduli lose their physical meaning, but data obtained from this test are still relevant because they represent real processing conditions. In particular, one can vary the strain amplitude and frequency in an independent manner. As a result, one can define additional strain- and frequency-dependent moduli which reduce to G' and G'' in the limit of small strain amplitudes [3].

The variation, with the rate of deformation, temperature, and other variables, of the different material functions defined in this section is discussed in detail in later chapters for a large variety of polymeric fluids and foods.

III. EXTENSIONAL FLOW

Besides viscometric flow, the other major category of flow of practical interest that can be generated in the laboratory is extensional flow. In mathematical terms, *extensional flow* may be represented in a rectangular Cartesian coordinate system x_i by the following set of equations for the three components of the velocity vector:

$$v_1 = \dot{\varepsilon}_1 x_1 \tag{1.9}$$

$$v_2 = \dot{\varepsilon}_2 x_2 \tag{1.10}$$

$$v_3 = \dot{\varepsilon}_3 x_3 \tag{1.11}$$

which also defines the *stretch rates* $\dot{\varepsilon}_i$. In uniaxial extension at constant stretch rate (see Figure 1.5), $\dot{\varepsilon}_1 = \dot{\varepsilon}$ and $\dot{\varepsilon}_2 = \dot{\varepsilon}_3 = -\dot{\varepsilon}/2$. The distance between material planes that are perpendicular to the flow direction increases with time in extensional flow. A material plane is a surface that always contains the same material points or particles. As material planes move apart, polymer molecules tend to uncoil, and ultimately one can even have stretching of chemical bonds which results in chain scission. Stresses in the flow direction can, therefore, reach fairly large values. By integrating Eq. (1.9), one finds that, at constant stretch rate, the length of a strip of material increases exponentially:

$$\text{In}\frac{L}{L_0} = \dot{\varepsilon}t \tag{1.12}$$

in which L_0 is the initial length and L is the length at some later time t.

The only stress that one can measure experimentally in uniaxial extension is the tensile stress σ_E, which, with reference to Figure 1.5, is

$$\sigma_E = \frac{F}{A} \tag{1.13}$$

For constant-stretch-rate homogeneous deformation that begins from rest, one defines a tensile stress growth coefficient

$$\eta_E^+(t, \dot{\varepsilon}) = \frac{\sigma_E}{\dot{\varepsilon}} \tag{1.14}$$

which has the dimensions of viscosity. The limiting value of η_E^+ as time approaches infinity is termed the *tensile* or *extensional* or *elongational viscosity*, η_E. In general, η_E is a function of the stretch rate, although in the limit of vanishingly low stretch rates [4]

$$\lim_{\dot{\varepsilon}\to 0}\frac{\eta_E}{\eta_0} = 3 \tag{1.15}$$

where η_0 is the *zero-shear-rate* viscosity.

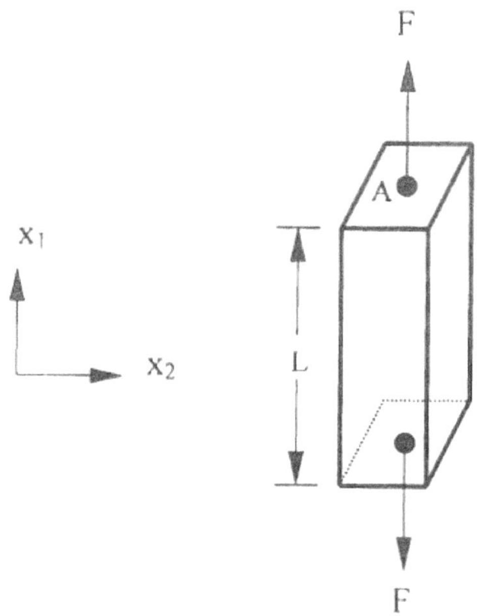

FIGURE 1.5 Schematic diagram for the measurement of extensional, or tensile, viscosity. (From Ref. [1].)

For Newtonian liquids, the extensional viscosity is three times the shear viscosity; but for polymeric liquids, the tensile viscosity may exceed the zero-shear viscosity by a couple of orders of magnitude. Extensional viscosity is of great practical importance when polymers and foods flow through channels or tubes in which the cross-sectional area is decreasing. Examples include flow through porous media, the spinning of fibers, the filling of molds in injection molding, and the mixing and baking of bread.

In addition to uniaxial extension, one also encounters biaxial extensional flow, especially equal biaxial extensional flow. In this case, $\dot{\varepsilon}_1 = \dot{\varepsilon}_2 = \dot{\varepsilon}_B$, and $\dot{\varepsilon}_3 = -2\dot{\varepsilon}_B$. Now, the measured stress in either stretching direction is σ_B, so that $\sigma_B = \sigma_{11} - \sigma_{33} = \sigma_{22} - \sigma_{33}$. One defines a biaxial stress growth coefficient η_B^+ as $\sigma_B(t)/\dot{\varepsilon}_B$, and the limiting value of this quantity as time approaches infinity is the *biaxial extensional* viscosity, $\eta_B(\dot{\varepsilon}_B)$. Another special case of biaxial extensional flow is *pure shear* where $\dot{\varepsilon}_1 = -\dot{\varepsilon}_2$, and $\dot{\varepsilon}_3 = 0$; this is less commonly encountered.

IV. THE STRESS TENSOR

If one isolates a rectangular parallelepiped of material having infinitesimal dimensions such as shown in Figure 1.6, one finds that two kinds of forces act on the material element. These are body forces and surface forces. *Body forces* result from the action of external fields such as gravity; *surface forces* express the influence

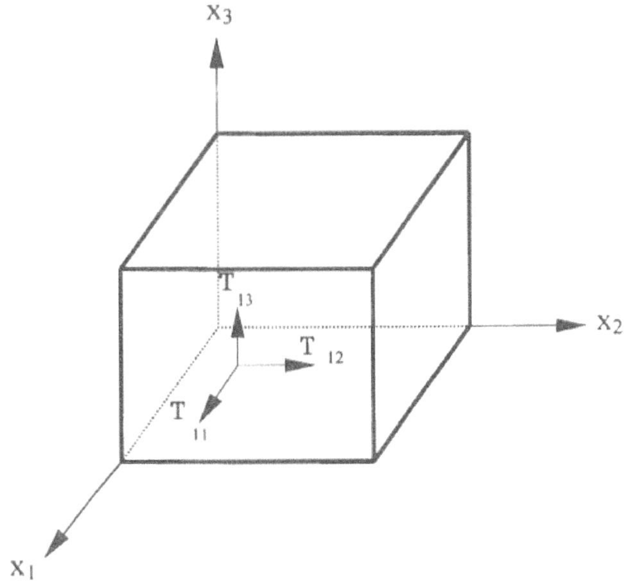

FIGURE 1.6 Diagram showing the notation used for components of the stress tensor. (From Ref. [5].)

of material outside the parallelepiped but adjacent to a given surface. Dividing the surface force by the area on which it acts yields the *stress vector*. Since the parallelepiped has six surfaces, there are six stress vectors.

Because each of the stress vectors can be resolved into three components parallel to the three coordinate axes, one has a total of 18 components. These are labeled T_{ij} or σ_{ij}, where the two subscripts help to identify a specific component. The first subscript, i, identifies the surface on which the stress acts; the surface, in turn, is identified by the direction of the outward-drawn normal. The second subscript, j, identifies the direction in which the stress component acts. Nine of the 18 components can be represented using a 3 × 3 matrix, called the *stress tensor:*

$$\begin{bmatrix} T_{11} & T_{12} & T_{13} \\ T_{21} & T_{22} & T_{23} \\ T_{31} & T_{32} & T_{33} \end{bmatrix}$$

T_{11}, for example, acts on a surface whose outward drawn normal points in the positive x_1 direction; the stress component itself also acts in the same direction. The other nine components of the stress are the same as these nine, but they act on opposite faces.

By means of a torque balance on a cubic element, it can be shown (see Ref. [2], for example) that T_{ij} equals T_{ji}. Thus, the stress tensor is symmetrical and only six of the nine components are independent quantities. Further, by examining the equilibrium

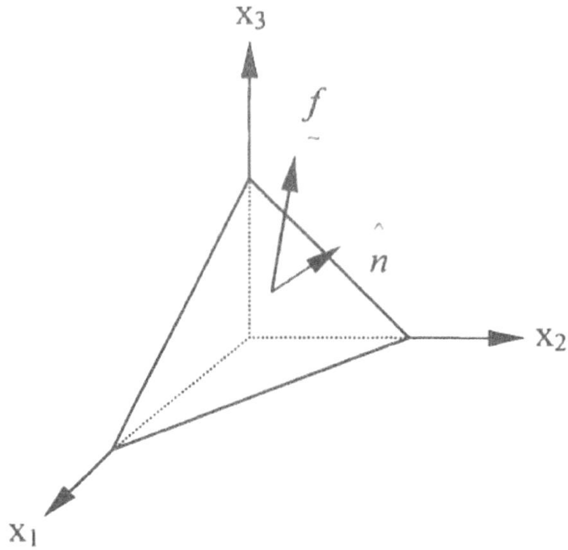

FIGURE 1.7 Equilibrium of a tetrahedron. (From Ref. [5]).

of the tetrahedron shown in Figure 1.7, it can be demonstrated [2] that if the normal to the inclined surface is $\hat{n}$ the components of the surface stress $\underset{\sim}{f}$ acting on that surface, f_i are (in matrix notation):

$$\begin{bmatrix} f_1 \\ f_2 \\ f_3 \end{bmatrix} = \begin{bmatrix} T_{11} & T_{12} & T_{13} \\ T_{12} & T_{22} & T_{23} \\ T_{13} & T_{23} & T_{33} \end{bmatrix} \begin{bmatrix} n_1 \\ n_2 \\ n_3 \end{bmatrix} \tag{1.16}$$

where n_i are the components of $\hat{n}$. Knowing the components of the stress tensor, therefore, allows one to obtain the stress vector acting on any plane described by the unit normal $\hat{n}$.

It is one of the major goals of rheologists to try to relate the stress tensor to the strain or the rate of strain. Recall that in Section II, it was shown that T_{12} is a unique function of the shear rate, while in Section III the net tensile stress was related to the stretch rate; the net tensile stress is nothing but T_{11}–T_{22}. These relationships between specific stress components and specific rate-of-strain components are known as *material functions*. The general equation relating the stress tensor to the three-dimensional strain or rate-of-strain tensor (the word *tensor* is simply another term for *matrix)* is known as a *constitutive equation*.

Due to fluid incompressibility, though, one cannot relate T_{ij} to deformation. This is because no amount of pushing, i.e., the application of hydrostatic pressure, can cause a change in volume. It is only when pressures are unequal that a strain, which can be understood as a change in the distance between neighboring particles, takes place. It is for this reason that it is usual to separate the stress tensor into two parts:

$$\begin{bmatrix} T_{11} & T_{12} & T_{13} \\ T_{12} & T_{22} & T_{23} \\ T_{13} & T_{23} & T_{33} \end{bmatrix} = \begin{bmatrix} -p & 0 & 0 \\ 0 & -p & 0 \\ 0 & 0 & -p \end{bmatrix} + \begin{bmatrix} \tau_{11} & \tau_{12} & \tau_{13} \\ \tau_{12} & \tau_{22} & \tau_{23} \\ \tau_{13} & \tau_{23} & \tau_{33} \end{bmatrix} \tag{1.17}$$

where p is the hydrostatic pressure whose presence causes no strain for incompressible materials and the τ_{ij} are components of the *extra stress* tensor whose presence causes strain to take place. Strain, therefore, is related to the extra stress $\underset{\approx}{\tau}$ rather than to the *total stress* $\underset{\approx}{T}$.

V. THE RATE-OF-STRAIN TENSOR

The shear rate defined in Eq. (1.2) can be generalized to three dimensions by defining the *rate-of-strain* tensor $\underset{\approx}{\dot{\gamma}}$ as

$$\dot{\gamma}_{ij} = \left(\frac{\partial v_i}{\partial x_j} + \frac{\partial v_j}{\partial x_i} \right) \tag{1.18}$$

and it is evident that for the flow situation in Figure 1.2, the shear rate is simply the 1,2 component of $\underset{\approx}{\dot{\gamma}}$. Furthermore, for a Newtonian liquid,

$$T_{ij} = \tau_{ij} = \eta \dot{\gamma}_{ij} \tag{1.19}$$

provided $i \neq j$, and

$$T_{ii} = -p + \eta \dot{\gamma}_{ii} \tag{1.20}$$

in which both i and j can be 1, 2, or 3.

Everything that we have defined so far has been with reference to a rectangular Cartesian coordinate system. In numerous practical situations, though, it is more appropriate to work in curvilinear coordinates. This parallel treatment may be found in more advanced books on rheology (see Ref. [6], for example).

Finally, we remark that the development of constitutive equations or, equivalently, the formulation of expressions similar to Eqs. (1.19) and (1.20) but applicable to polymeric fluids is merely a means to an end. In modeling polymer processing operations, one needs to solve boundary value problems involving the flow of polymeric fluids. The solution procedure involves inserting the constitutive equation into the momentum and energy balance equations and solving the resulting coupled differential equations subject to appropriate boundary conditions. In this regard, the no-slip boundary condition, which states that at a fluid–solid boundary the velocity of the fluid is the same as that of the solid, has been an article of faith for an extremely long time. Evidence [7], though, suggests that this boundary condition may not be universally valid. Indeed, slip at the solid–liquid boundary has been hypothesized to be the cause of polymer processing instabilities such as melt fracture; these instabilities tend to limit the rate of production during processes such as extrusion, fiber spinning, and film blowing [8].

Additionally, the complex structures and flow behavior of food and polymeric fluids during industrial processing have continued to elude the choice of a universal constitutive equation, in spite of the formulation of many constitutive equations [9, 10].

REFERENCES

1. L. E. Nielsen. Polymer Rheology. Marcel Dekker, New York, 1977.
2. I. H. Shames. Mechanics of Fluids. 4th ed., McGraw Hill, New York, 2002.
3. M. Kamkar, R. Salehiyan, T. B. Goudoulas, M. Abbasi, C. Saengow, E. Erfanian, S. Sadhegi, G. Natale, S. A. Rogers, A. J. Giacomin, and U. Sundararaj. Large amplitude oscillatory shear flow: Microstructural assessment of polymeric systems. Prog. Polym. Sci. 132: 101580 (2022).
4. K. Walters. Rheometry. Chapman and Hall, London, 1975.
5. R. K. Gupta. Polymer and Composite Rheology. 2nd ed., Marcel Dekker, New York, 2000.
6. R. B. Bird, R. C. Armstrong, and O. Hassager. Dynamics of Polymeric Liquids. 2nd ed. Vol. I, Wiley, New York, 1987.
7. A. V. Ramamurthy. Wall slip in viscous fluids and influence of materials of construction. J. Rheol. 30:337–357 (1986).
8. C. J. S. Petrie and M. M. Denn. Instabilities in polymer processing. A.I.Ch.E J. 22:209–236 (1976).
9. R. G. Larson. Constitutive Equations for Polymer Melts and Solutions. Butterworths, Boston, 1988.
10. E. Narimissa and M. F. Wagner. Review on tube model based constitutive equations for polydisperse linear and long-chain branched polymer melts. J. Rheol. 63:361–375 (2019).

2 Instruments for Shear Rheology

I. INTRODUCTION

A great variety of instruments have been used to measure the rheological properties of polymer solutions and polymer melts when these are subjected to shear deformation [1–5]. These instruments range in sophistication from simple melt indexers employed for quality control purposes to very complicated, research-grade rheometers capable of measuring, as a function of temperature, pressure, and rate of deformation, both steady-state and transient values of the various material functions defined in the previous chapter. As might be expected, these instruments span a wide price range [2, 3], depending on the capability of the viscometer, and they are manufactured and marketed by a large number of manufacturers all over the world. In this chapter, we consider general-purpose, shear instruments that one might expect to find in any laboratory devoted to rheological work; these include capillary viscometers as well as rotational viscometers. We describe the essential features of each type of instrument, present the basic equations used for data analysis, and discuss equipment limitations and sources of error. A consideration of actual data, their possible molecular interpretation, and representation using constitutive equations is deferred to later chapters. Extensional viscometers are examined in Chapter 7, while specialized instruments are introduced throughout this book as appropriate.

Since shear viscosity has been measured for a very long time, we might expect shear rheometry to be a mature field. However, new instruments are continually being introduced. Some, such as the sliding-plate rheometer, involve new concepts; others improve on existing instruments as far as the ease of measurement and data analysis or the range of measurement is concerned. Yet others allow us to make measurements on complex materials or materials that might be reactive or temperature sensitive.

II. CAPILLARY VISCOMETERS

The standard instrument for the measurement of the steady-shear viscosity of polymeric fluids and foods at shear rates witnessed in typical processing operations is the *capillary viscometer*, shown schematically in Figure 2.1. In this device, the sample is forced by a piston or by gas pressure from a gas reservoir through a capillary; to obtain consistent results, the ratio of the reservoir diameter to the capillary diameter has to exceed about 12. In commercial instruments, the capillary is usually vertical, and it can be detached from the viscometer so that we may make measurements with other capillaries having different entrance angles or different values of length L or diameter D. In actual use, we fill the cold reservoir, called the *barrel*, with a

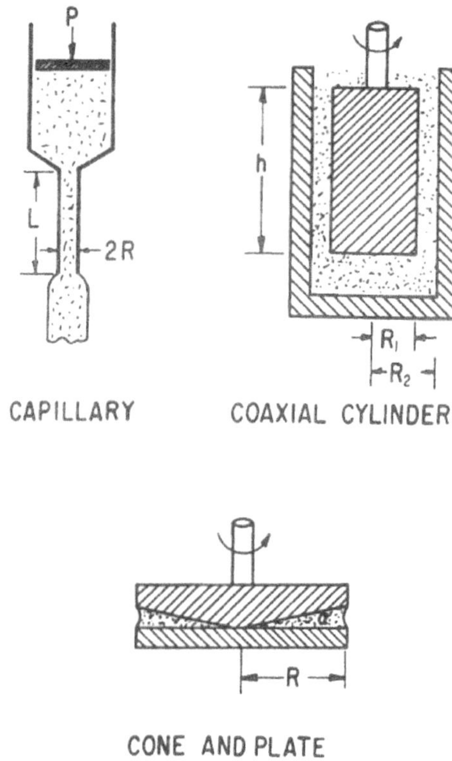

FIGURE 2.1 Schematic diagrams of three types of viscometers. (From Ref. [6].)

fluid or with pellets of the neat or filled polymer and then heat the system to the test temperature. Once thermal equilibrium is attained, extrusion is begun, and the volumetric flow rate Q of polymer coming from the capillary at a given pressure drop is recorded. If the plunger speed can be specified, as is commonly the case, it is not necessary actually to collect the extrudate. A load cell attached to the piston or a pressure transducer attached near the barrel exit is employed to measure the pressure of extrusion.

The capillary rheometer has a number of advantages. First, the instrument is relatively easy to fill. This is an important consideration, since most polymer melts are too viscous to pour readily even at high temperatures. The test temperature and rate of shear are varied readily. The shear rates and flow geometry are similar to the conditions actually found in pipe flows, extrusion, and in injection molding. In addition to the viscosity, some indication of polymer elasticity is found from the die swell of the extrudate. Finally, factors affecting the surface texture of the extrudate and the phenomenon of melt fracture can be studied. However, there are disadvantages too. The principal disadvantage is that the rate of shear is not constant but varies with radial position within the capillary. Another disadvantage is the necessity of making a number of corrections in order to get accurate viscosity values, especially with low-viscosity materials or when short capillaries are used.

In this flow field, known as circular *Poiseuille flow*, the velocity profile in the capillary is fully developed only in the region away from the inlet and the exit. At the capillary ends, there must be velocity rearrangements, and the pressure gradient varies with axial position and is larger than that in the fully developed region. If, however, the ratio of the capillary length to the capillary diameter is large, say, 50 or more, we may neglect end effects and assume that the pressure gradient in the entire capillary is $(p_1 - p_2)/L$; this is the difference between the inlet and exit pressures divided by the capillary length. By means of a force balance, then, the magnitude of the shear stress at the capillary wall, τ_w, is as follows:

$$\tau_w = (p_1 - p_2)\frac{R}{2L} \tag{2.1}$$

where R is the capillary radius. To determine the shear viscosity, we need to divide the wall shear stress by the shear rate at the wall, $\dot{\gamma}_w$. For any non-Newtonian fluid, the expression for the wall shear rate is given as [7]:

$$\dot{\gamma}_w = \frac{4Q}{\pi R^3}\left(\frac{3}{4} + \frac{1}{4}\frac{d\ln Q}{d\ln \tau_w}\right) \tag{2.2}$$

To measure the viscosity as a function of shear rate, the flow rate is determined as a function of the pressure drop. This can be converted to flow rate as a function of the wall shear stress using Eq. (2.1), which then allows us to compute the derivative appearing in Eq. (2.2) at any given value of the wall shear stress. The ratio of this wall shear stress to the corresponding wall shear rate calculated from Eq. (2.2) gives the shear viscosity at one shear rate—the one corresponding to the chosen value of the shear stress at the wall. By repeating this process at other values of the wall shear stress, we can trace out the entire η-versus-$\dot{\gamma}$ curve or the *flow curve*.

If the non-Newtonian fluid obeys the power law (see Chapter 3), in which the shear stress depends upon the shear rate to the nth power, Eq. (2.2) takes the form:

$$\dot{\gamma}_w = \frac{4Q}{\pi R^3}\left(\frac{3n+1}{4n}\right) \tag{2.3}$$

where

$$n = \frac{d\log\tau}{d\log\dot{\gamma}} \tag{2.4}$$

In the special case of a Newtonian liquid, n equals unity and the wall shear rate is simply $4Q/\pi R^3$. It is, therefore, seen that the quantity in parentheses in Eq. (2.2) corrects the rate of shear for non-Newtonian liquids [8]; this is known as the *Rabinowitsch correction* and $4Q/\pi R^3$ is called the *apparent wall shear rate*. Clearly, for a Newtonian liquid, the shear viscosity is as follows:

$$\eta = \frac{\pi R^4 (p_1 - p_2)}{8LQ} \tag{2.5}$$

The use of Eq. (2.5), which is the well-known *Hagen–Poiseuille equation*, allows us to use Newtonian liquids to calibrate the viscometer and to check that the experimental procedure is satisfactory. Note that the shear rate is always a maximum at the wall but that the velocity is zero at the wall. The characteristics of capillary flow are shown in the bottom section of Figure 2.2. The solid lines are for Newtonian liquids; the dashed lines are typical of polymeric non-Newtonian fluids. Under conditions in which Q is the same for both Newtonian and non-Newtonian fluids, the area under the velocity distribution curves should be the same rather than as shown in Figure 2.2, where conditions have been adjusted to match the velocities at the center of the capillary. Note the validity of Eqs. (2.1)–(2.5) rests on several assumptions. These include constant temperature conditions, no slip at the capillary wall, and no effect of pressure on either the density or viscosity of the fluid.

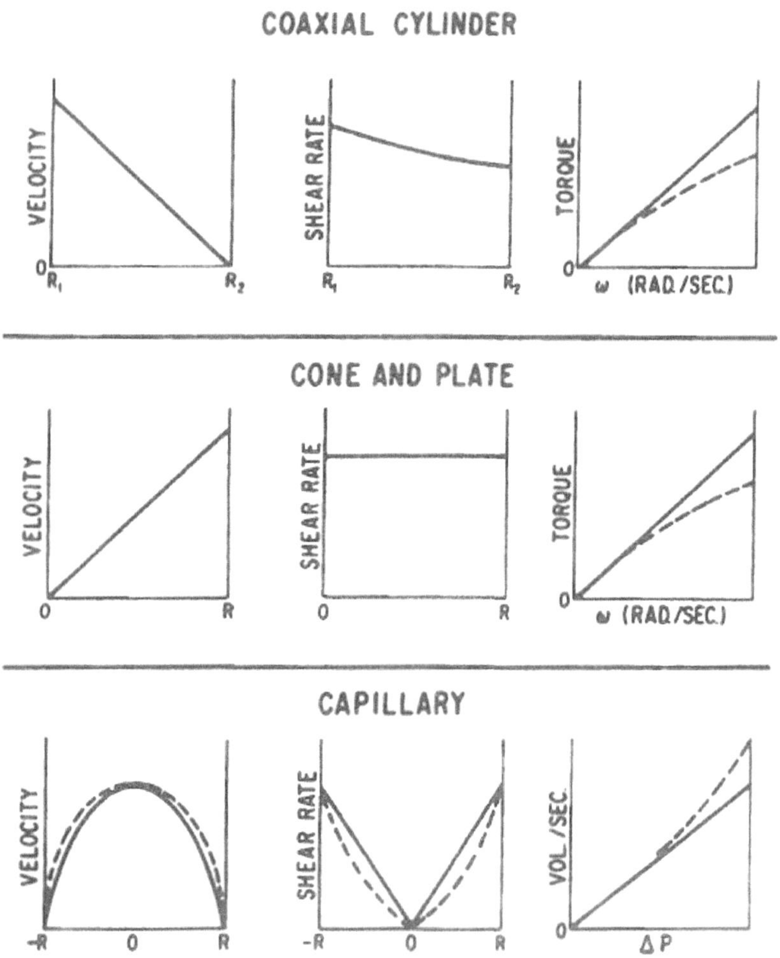

FIGURE 2.2 Characteristics of the three viscometers shown in Figure 2.1. Solid lines refer to Newtonian fluids; broken lines are typical of non-Newtonian polymer melts. (From Ref. [6].)

A small-diameter capillary is preferred in order that one may make measurements using small-sized samples (working volumes of 25 ml are common). Even for neat polymer melts, the practical lower limit is about 0.025 cm; otherwise, blockage can occur due to the presence of impurities in the polymer. In the case of foods, even neat systems such as starches do not melt readily and tend to remain heterogeneous requiring the use of larger diameter capillaries; smaller capillaries may be used for solutions of hydrocolloids and proteins. Furthermore, the use of capillaries having large *L*/*D* values is desirable for theoretical reasons. However, this is possible only for low-viscosity liquids for which the pressure drop would otherwise be small. When working with molten polymers, long capillaries cannot be used due to thermal degradation problems and because the pressure drops are excessive; the latter fact causes problems with data analysis, since the viscosity can depend on pressure at high pressures. Excessive pressures can also be limiting for highly viscous foods such as dough.

If short capillaries are used, the variation in the pressure gradient along the tube axis cannot be neglected, and the effective length of the capillary becomes greater than the true length. This is taken into account through the Bagley correction [9, 10]. The shear stress at the wall of Eq. (2.1) becomes

$$\tau_w = \frac{R(p_1 - p_2)}{2(L + eR)} = \frac{p_1 - p_2}{2(L/R + e)} = \frac{(p_1 - p_2) - p_0}{2L/R} \tag{2.6}$$

The *Bagley correction* factor e should be independent of capillary length, but in general it does vary somewhat with L/R because of the elasticity of polymer melts. The Bagley correction is determined by measuring the pressure drop at constant rate of shear for several capillary lengths and extrapolating to zero pressure drop, as shown in Figure 2.3. In Eq. (2.6), p_0 is the pressure drop corresponding to a capillary of zero length, an orifice, for a given rate of shear.

The upper limit on the shear rate in capillary viscometers is about 10^5–$10^6 s^{-1}$. This comes about for a variety of reasons. As mentioned earlier, these instruments come in two basic designs, which differ in the method of melt extrusion. In the pressure-driven instrument, an inert gas, such as nitrogen, is used to pump the liquid out of the reservoir, and we measure the volumetric flow rate corresponding to the applied pressure drop; on-line flow rate determination is possible, and practical details are available in [11–13]. For a capillary of given length and diameter, the maximum shear rate is set by the maximum pressure, but it can be increased further by increasing the capillary diameter. As seen from Eq. (2.1), the wall shear stress is directly proportional to the tube radius, and, if the shear stress increases, the shear rate must increase as well. Increasing the diameter, though, causes a rather disproportionate increase in the volumetric flow rate, which can also lead to nonisothermal conditions due to viscous heating [14, 15]. In a plunger-driven instrument, the extrusion velocity is fixed, and the corresponding pressure drop is measured. Now the maximum shear rate is decided by the maximum speed of the plunger, but it can be increased further by reducing the capillary diameter. This is evident from an examination of Eq. (2.2). However, we cannot exceed the upper limit of the force transducer; for instruments designed to work with filled polymers, this force limit may be as high

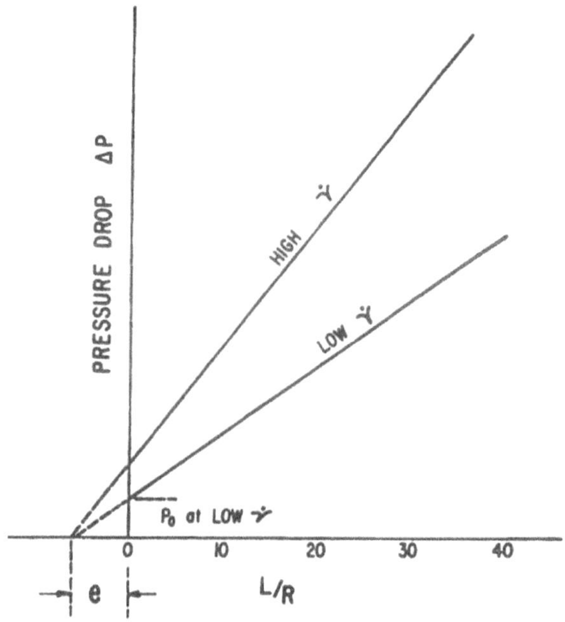

FIGURE 2.3 Bagley correction for capillary rheometers. (From Ref. [6].)

as 100 kN. In addition, data cannot be trusted once instabilities such as melt fracture set in. The lower limit of operation of capillary viscometers is about 1–10 s^{-1}. This is determined by the occurrence of drips in a pressure-driven instrument, while for a plunger-type viscometer the lower limit is set by the amount of friction between the plunger and the barrel. As the plunger speed decreases, friction can become a significant part of the total pressure drop and make the viscosity measurements unreliable. Another point of concern is the possibility that the sample may slip at the capillary wall, especially in the presence of processing aids [13]. This topic is addressed in detail in volume 2, and more details about capillary viscometry are available in References [1–5, 16]. The development of a capillary viscometer capable of operating at pressures substantially above atmospheric has been described by Galvin et al. [17], and a variety of instruments are now available for this purpose. Details of these viscometers have been reviewed, and it seems that a capillary device may be the instrument of choice [18, 19]. In this regard, the multi-pass rheometer which consists of a capillary held between two barrels connected to two movable pistons that allow for the imposition of a hydrostatic pressure has received significant attention [20].

Almost all present-day commercial instruments are plunger-driven and are quite versatile. Some, such as the Rosand RH7 and RH10 series viscometers from Netzsch, can operate at temperatures upto 500°C, can have twin capillaries that allow two simultaneous experiments, and have the facility of a nitrogen purge for work with materials that are sensitive to moisture and oxygen. In addition, some viscometers have pressure transducers mounted near the capillary entrance. This allows us to obtain a more accurate estimate of the pressure drop, since we do not have to account for pressure losses in the barrel. An advantage of the twin-bore design, with pressure transducers, is that simultaneous measurements of the pressure drops on orifice and long capillaries of the same diameter can be conducted; the orifice pressure drop (p_0 in Eq. 2.6) can then be used to perform the Bagley correction directly. Often these viscometers are used as extruders in laboratory experiments, and it is common to employ them for spinning fibers and for determining the melt strength of polymers.

In contrast to the sophisticated capillary viscometers described above, a simple capillary instrument, called the *melt indexer*, is found in almost all industrial laboratories. Here, one characterizes polymer melts through a parameter called the *melt flow index* (MFI). This is the mass of polymer in grams extruded in 10 min through a capillary of specified length and diameter by the application of pressure using a dead weight as specified by ASTM D 1238. For polyethylene, the load is 2.16 kg, and the capillary diameter and length are, respectively, 2.095 and 8 mm; the temperature used is 190°C [21]. The MFI, therefore, is a single-point viscosity measurement at relatively low shear rate and temperature. One can also use a 10 kg load to obtain the *high-load melt* index, and two data points allow for the determination of the power-law parameters [21]. Shenoy et al. [22] have shown that temperature-independent viscosity master curves can be generated for a given polymer type by plotting the product of the shear viscosity and MFI against the ratio of the shear rate to MFI. They have also provided master curves for several commodity thermoplastics [23], engineering plastics [24], and specialty polymers [25].

Although it is a fairly routine matter to use a capillary viscometer to measure the rheological properties of polymer solutions and polymer melts, suspensions and emulsions have to be treated with care. For both suspensions and emulsions, data analysis can be complicated due to (1) structural changes that occur during the process of measurement, and (2) apparent slip at the viscometer wall due to depletion of the dispersed phase in the vicinity of a solid surface. These effects are considered in later chapters. A note of caution also needs to be sounded on the practical side. Highly loaded suspensions are known to be shear thickening, and the sudden increase in the shear stress can make the viscometer seize, resulting in physical harm to the instrument and the investigator.

In closing this section, we note that the theory of Poiseuille flow through slits, called *plane Poiseuille flow*, is similar to that for circular tubes [7], and *slit rheometers* are also commercially available [2, 3], including as an accessory to commercial capillary rheometers. With two or more pressure transducers along the length, the wall the shear stress can be determined directly without the need for the estimation of the Bagley correction [11]. An advantage of slit rheometry is that shear rates greater than 10^6 s^{-1} can be attained and isothermal conditions still maintained. As shown by Lodge in Collyer and Clegg [3], this feature can be used to measure normal stress differences (see Chapter 5) at shear rates higher than those accessible with rotational instruments. Both capillary and slit viscometers find application in industry as "on-line" process instruments for monitoring rheological properties of foods and polymer melts during extrusion [26–28]. A gear pump is used to withdraw a sample stream that is then made to go through a slit process rheometer fitted with pressure transducers. Once the viscosity measurement is made, the side stream can be returned to the extruder by means of a second gear pump. *On-line viscometers* are made by several companies, and in the Thermo Scientific HAAKE PCR 630 instrument, for example, one can measure the melt index as well as the shear viscosity. This viscometer uses a slit die ranging in length from 4 to 8 cm, in height from 0.03 to 0.2 cm, and in width from 1.2 to 2.25 cm. The maximum shear rate possible is 4000 s^{-1}, and the maximum possible operating temperature is 400°C.

III. COAXIAL CYLINDER VISCOMETERS

The coaxial cylinder, or concentric cylinder, or *Couette viscometer*, is frequently used for measuring the steady-shear viscosity and dynamic mechanical properties of low-viscosity liquids, polymer solutions, solid-in-liquid suspensions, and emulsions. However, it is used only occasionally at high temperatures or for high-viscosity melts. In this instrument, as shown in Figure 2.1, the fluid sample is placed in the annular space between two concentric cylinders. In one version of the apparatus, in order to measure the liquid viscosity, either one of the two cylinders is rotated at a constant speed while the torque acting on the other (stationary) cylinder is measured. In another version of the viscometer, one cylinder is held stationary while the other one is rotated under the influence of a constant torque; the speed of rotation is measured by some kind of a tachometer device. When it is desired to measure dynamic properties, the constant speed of rotation is replaced by oscillation at a fixed frequency and small amplitude. The discussion below will focus on instruments that rotate the outer cylinder and measure the torque through the inner cylinder; note that these instrument designs are generally referred to as Separate Motor and Transducer.

A major advantage of the coaxial cylinder viscometer is the nearly constant shear rate throughout the entire volume of fluid being measured if the gap between cylinders is small; this criterion is satisfied by ensuring that the ratio of the diameter of the inner cylinder to the diameter of the outer cylinder exceeds 0.97. This is an important factor with non-Newtonian fluids, in which the viscosity may be strongly dependent upon the rate of shear. Coaxial cylinder instruments are simple to use and are calibrated easily, and the corrections can be small. Most of these instruments are able to go to very low deformation rates, which is an advantage when working with blood and biological fluids, where one does not want structure alteration during the process of measurement. A major disadvantage of these instruments is the difficulty in filling them with a very viscous polymer melt. A less important, but annoying, disadvantage is the creeping of the polymer up the shaft of the inner cylinder because of normal stresses developed in the polymer by the rotation of the cylinder.

For Newtonian liquids, the important equations for coaxial cylinder viscometers (neglecting end corrections) are given below [29]. The corresponding results for power-law fluids may be found in [12].

$$\tau(r) = \frac{M}{2\pi r^2 h}; \quad R_1 \le r \le R_2 \tag{2.7}$$

$$\dot{\gamma}(r) = \frac{2\omega R_1^2 R_2^2}{r^2\left(R_2^2 - R_1^2\right)} \tag{2.8}$$

A torque M is produced by the rate of angular rotation ω in radians per second. As shown in Figure 2.1, the radii of the inner and outer cylinders are R_1 and R_2, respectively. The immersed length of cylinder is h, and r is any arbitrary radius between R_1 and R_2. The top part of Figure 2.2 illustrates the characteristics of coaxial cylinder instruments. The solid curves are for Newtonian liquids; the dashed line is typical of a non-Newtonian liquid. Note that Eq. (2.7), the relationship between the shear stress and the torque, is independent of the constitutive nature of the fluid. If the annular spacing is small compared to the diameter of the cylinders, the shear rate is nearly constant across the annular gap for all fluids; this shear rate is given by:

$$\dot{\gamma} = \frac{\omega R_1}{R_2 - R_1} \tag{2.9}$$

The main correction that must be applied to data in order to get the correct viscosity is an end correction for a coaxial cylinder viscometer. The liquid at the bottom of the inner cylinder creates additional drag and gives rise to a torque in addition to the torque due to the liquid in the annular gap. Thus, the measured torque is equivalent to that of an inner cylinder with an apparent length greater than its true length. The simplest way to account for this additional length is to calibrate the viscometer with a liquid of known viscosity and to use the following equation:

$$\eta = \frac{kM}{\omega} \tag{2.10}$$

The form of Eq. (2.10) follows from the fact that viscosity is the ratio of the shear stress to the shear rate, and these two quantities are given, respectively, by the right-hand sides of Eqs. (2.7) and (2.9). The instrument constant k takes into account any corrections as long as the volume of liquid in the viscometer is held constant. Of course, the end correction is negligible if the immersed length is a couple of orders of magnitude larger than the annular gap. The correction can also be minimized by shaping the bottom of the inner cylinder to resemble an inverted cone, creating a recessed bottom.

The maximum value of the shear rate achievable with coaxial cylinder viscometers is usually not an instrument limitation but depends on the nature of the fluid sample. For high-viscosity liquids, viscous heating may become a problem; for low-viscosity liquids, the upper limit may be set by the occurrence of secondary flows or the transition to turbulent flow. Usually, though, we can make measurements at shear rates up to about 100 s^{-1}. At the other end of the scale, one can often go down to shear rates as low as 10^{-2} s^{-1}, especially with high-viscosity liquids; the limiting factor here is the sensitivity of the torque transducer or the torsion bar or spring employed to measure the torque. Double gap concentric cylinders can be useful for low-viscosity liquids and low shear rates. In the Contraves low shear instruments available earlier, one was able to measure torque values as low as 10^{-9} Nm by replacing the torsion bar with a light-deflection compensating system. Here, the outer cylinder or cup is driven by a motor and the angular deflection of the stationary inner cylinder or bob is detected by a light beam reflected from a mirror attached to the bob axis and monitored by a photocell. In most coaxial cylinder viscometers, the sample size is about 1 ml and the maximum temperature is less than 80°C.

When the coaxial cylinder viscometer is used in the oscillatory mode, one cylinder is rigidly supported while the other one is given a small-amplitude angular displacement of the form:

$$\phi = \phi_0 \sin(\omega t) \tag{2.11}$$

Now the torque becomes time-dependent, but the shear stress (and this is sinusoidal) is still related to the torque by Eq. (2.7). The sinusoidal strain is easily determined from the geometry and is given by:

$$\gamma(t) = \frac{R_2 \phi}{R_2 - R_1} \tag{2.12}$$

A knowledge of the stress and strain as a function of time allows us to calculate the complex viscosity with the help of Eqs. (1.6)–(1.8). If, however, one of the cylinders is not rigidly held but is suspended from a torsion wire so that it too can oscillate, the theoretical analysis is somewhat involved, and the relevant results may be found in Walters [7].

The coaxial cylinders is a popular geometry used in the rheological measurements of less viscous foods. To account for the complex rheology of foods, such as heterogeneous and sensitive structures, the inner cylinder can be replaced with other geometries such as a vane or helical impeller. These may improve the precision of the measurement at the expense of accuracy. Such variants are offered by the major manufacturers of rotational rheometers as off-the-shelf or custom geometries.

A variation of the coaxial cylinder viscometer is the "*Brookfield*" dial viscometer—a cylindrical spindle rotating in a 600-ml container of fluid. While the AMETEK Brookfield Company makes a variety of rotational viscometers, it is this simple, inexpensive device, commonly used for quality control purposes, that has become synonymous with the name of the company. The spindle is driven by a synchronous motor through a calibrated spring; the deflection of the spring is indicated by a pointer that displays a *"dial" viscosity*. Recent versions of these viscometers come with a digital display for the reading. When measurements are made on suspensions and pastes, the cylindrical spindle often cuts a hole in the material and the reading falls progressively. To prevent fluid channeling, we replace the cylindrical spindle with a T bar that is continuously raised and lowered in the fluid container. Again, the measured viscosity is not a fundamental quantity but an empirical measure of the resistance to deformation of the fluid.

IV. CONE-AND-PLATE VISCOMETERS

The geometrical characteristics of a *cone-and-plate rheometer* are shown in the bottom section of Figure 2.1. A flat, circular plate and a linearly concentric cone are rotated relative to each other. The cone is normally truncated so that there is no physical contact between the two members. The liquid is in the space between the plate and cone. The plate radius R is typically a couple of centimeters, whereas the cone angle α is usually a few degrees. Either of the two members can be rotated or oscillated, and we measure the torque M needed to keep the other member stationary. We also measure the downward force N needed to hold the apex of the truncated cone at the center of the disk. From these measurements we deduce the various material functions defined in Chapter 1. Note that N equals zero for a Newtonian liquid.

A major advantage of a cone-and-plate viscometer is the constant shear rate throughout all the liquid. This comes about because, at a fixed radial position, the circumferential velocity varies linearly across the gap between the cone and the plate. The constancy of shear rate is especially important, since it makes data analysis extremely straightforward. A consequence of this is that the cone-and-plate viscometer is well suited to making time-dependent measurements, such as dynamic measurements or transient measurements following a step change in the rate of shear strain. If disposable fixtures are employed, measurements can be made readily on chemically reactive systems, allowing for the investigation of resin curing and the phenomenon of gelation. This instrument also has other advantages. The sample size is very small, so the instrument is valuable for evaluating the rheological behavior of experimental samples, where only small quantities of material may be available. The small sample size also results in less heat buildup at high rates of shear. Another advantage is the ease of loading the sample and the ease of cleaning the apparatus at the end of a test. The instrument has a number of disadvantages: measurements can usually not be made at high shear rates. With increasing shear rate, there is a tendency for the development of secondary flows in the sample and for viscoelastic fluids, such as polymers, for the sample to crawl out of the gap under the influence of centrifugal forces and elastic instabilities. Even otherwise, the shape of the free surface at the edge of the sample can get distorted, leading to inaccuracies; this phenomenon can be aggravated by sample drying in the case of foods, polymer solutions, and suspensions. If this happens, coating the free surface with silicone oil or using a solvent trap can often provide relief. Accurate spacing of the cone-and-plate is required at all times, especially for the measurement of transient normal stresses. This makes the equipment more sophisticated and expensive, and it also demands a higher level of skill of the investigator.

For both Newtonian and non-Newtonian liquids, the basic equations for cone-and-plate rheometers are as follows [7]:

$$\tau = \frac{3M}{2\pi R^3} \tag{2.13}$$

$$\dot{\gamma} = \frac{\omega}{\alpha} \tag{2.14}$$

$$N_1 = \frac{2N}{\pi R^2} \tag{2.15}$$

$$T_{\theta\theta} = -p_a + N_2 + \left(N_1 + 2N_2\right)\ln\frac{r}{R} \tag{2.16}$$

where the cone angle is measured in radians, p_a is atmospheric pressure, and $T_{\theta\theta}$ is the total stress exerted by the fluid normal to the surface of the plate; the subscripts on T represent the θ direction in a spherical coordinate system.

The characteristics of the cone-and-plate viscometer are displayed in the middle portion of Figure 2.2. As before, the shear viscosity is obtained by dividing the shear stress by the shear rate. In principle, a knowledge of N_1 permits the determination of N_2 through the use of Eq. (2.16) provided that a pressure transducer is employed to measure $T_{\theta\theta}$ on the plate surface. This measurement, however, is not easy to make with any degree of accuracy. If the angular velocity is not constant but is some specified function of time—say, sinusoidal or a step function—the measured quantities M and N also depend on time but the preceding above equations remain unaltered. This makes it easy to compute dynamic mechanical properties or to measure stress growth and stress relaxation. Note that today's viscometers can be operated not just at particular values of the shear rate but also at specified values of the shear stress; the shear rate then becomes the dependent variable. The use of a constant-stress rheometer is one way of determining creep behavior and the value of the yield stress, if any, of a fluid at hand.

The cone-and-plate viscometer is the workhorse of the rheological testing laboratory, and it is manufactured by almost every company that makes rheometers. Depending on the capability of the torque transducer and the viscosity of the liquid sample, measurements can be made over a wide shear rate range, extending from 10^{-6} to 5000 s^{-1} (but recall the caution about secondary flows and the effect of centrifugal force). As a general rule, high viscosities and low shear rates go hand in hand, and the low-shear-rate capability makes it possible to determine the zero-shear-rate viscosity of polymer melts. For oscillatory experiments, a frequency range of 10^{-2}–100 rad/s is common. In the ARES rotational rheometer from TA Instruments, and this is one of the most sophisticated instruments available, the torque range is from 0.1 μN.m to 200 mN.m, and the normal force is measured in the range of 0.001–20 N. The sample is usually kept in an inert atmosphere in an oven that can be heated up to about 600°C or cooled down to about −150°C at heating and cooling rates as high as 60°C/min.

V. PARALLEL-PLATE VISCOMETERS

Even though the cone-and-plate viscometer is the preferred instrument when measurements are made at low shear rates, difficulties arise during measurements on filled polymers if the size of the particulates is comparable to the distance between the truncated cone and the surface of the plate. For short-fiber suspensions, however, Eberle et al. [30] employed "donut"-shaped samples. These were made by drilling a hole through the center of the normal samples employed with cone-and-plate fixtures. The more normal solution is shown in Figure 2.4a, where one replaces the cone by a circular plate so that one has torsional flow between two identical, coaxial, parallel disks; the rest of the apparatus remains unchanged. This is known as a *parallel-plate viscometer.* One again measures the torque M and the normal force N tending to separate the two plates when the gap between them is maintained fixed. If one uses a large gap, an advantage of this arrangement is that fairly large and measurable torque readings are obtained during small-amplitude oscillatory flow. Another advantage is that sample preparation is easy, since disks of the appropriate size can be cut or punched out of a flat sheet of material. With pastes and thick suspensions, as many foods are, the sample may be loaded by squeezing them between the plates

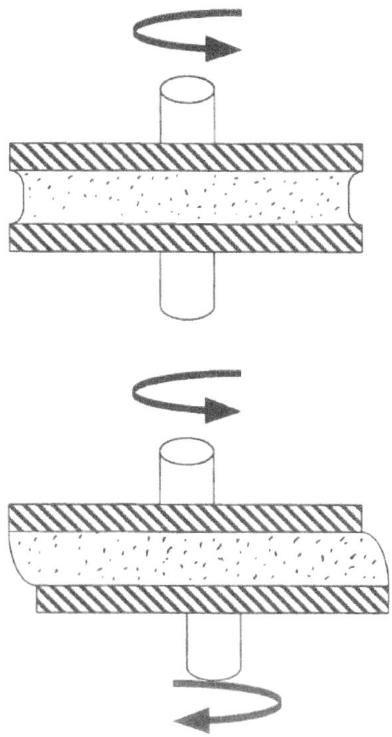

FIGURE 2.4 Schematic diagrams of (a) the parallel-plate rheometer and (b) the orthogonal rheometer. (From Ref. [33].)

to a desired gap; unlike cone-and-plate, one is not required to always achieve the exact same gap. The major disadvantage of this instrument is that the shear strain and the shear rate vary with radial position, complicating the process of data analysis. In steady-shear flow, the shear rate and shear stress at the edge of the disks located at $r = R$ are given by [7]:

$$\dot{\gamma}(R) = \frac{\omega R}{h} \tag{2.17}$$

$$\tau(R) = \frac{3M}{2\pi R^3}\left[1 + \frac{1}{3}\frac{d\ln M}{d\ln\dot{\gamma}(R)}\right] \tag{2.18}$$

$$N_1(R) - N_2(R) = \frac{2N}{\pi R^2}\left[1 + \frac{1}{2}\frac{d\ln N}{d\ln\dot{\gamma}(R)}\right] \tag{2.19}$$

in which h is the vertical gap between the two plates. It is clear that in order to get the shear stress or the normal stress differences at a single value of the shear rate, we have to measure the torque and normal force at several values of the shear rate and differentiate the data; data can be collected either by varying the gap spacing or, more easily, by changing the speed of rotation. Unfortunately, the two normal stress differences cannot be determined separately unless some very gross assumptions are made. However, if the first normal stress difference is measured using a cone-and-plate viscometer, the parallel-plate instrument can be utilized to compute the second normal stress difference. A variation of the experimental procedure already outlined [31] is to apply a known downward normal force to the upper plate and to measure the resulting value of h.

A single-point correction may be used to obtain the true viscosity and the normal stress coefficient, by avoiding the differentiation [32]. Most instruments report an *apparent* viscosity for the shear rate shown in Eq. (2.17). This viscosity becomes the true value at the lower shear rate of $0.755\times\dot{\gamma}(R)$. The single-point correction can provide rapid results for quality control laboratories. A similar correction can also be applied to the normal stress difference [32].

In small-amplitude sinusoidal deformation with an angular amplitude of ϕ_0 such that the entire sample is still within the linear viscoelastic region, a measurement of the torque amplitude M_0 and its phase difference δ with the angular deformation yields the following [2]:

$$G' = \frac{3M_0 h\cos\delta}{\pi R^4 \phi_0} \tag{2.20}$$

$$G'' = \frac{2M_0 h\sin\delta}{\pi R^4 \phi_0} \tag{2.21}$$

Instead of oscillating one plate relative to the other, the parallel-plate viscometer is sometimes operated in the "eccentric rotating disk" mode, as shown in Figure 2.4b. In this orthogonal rheometer, there is an offset between the axes of the two disks; as one disk rotates relative to the other, an eccentric oscillatory motion is set up in the polymer [34]. An analysis of this flow field [7] again yields G' and G''.

Equipment limitations and sources of error for the parallel-plate viscometer are essentially the same as those for the cone-and-plate viscometer. When working with suspensions and emulsions, a common problem encountered with both types of instruments is the tendency of the sample to slip at the fixture walls [35]; roughening the fixture surfaces or using serrated surfaces is the normal method of attempting to overcome this difficulty.

VI. SLIDING-PLATE VISCOMETERS

Polymer processing operations are generally characterized by large deformations of short duration during which a steady state in the stress is usually not achieved. Consequently, if the purpose of rheological testing is to examine polymer processability, most of the viscometers described in earlier sections of this chapter may be inadequate, since these are designed to measure steady-state stresses at relatively low rates of deformation. An instrument that is well suited to the investigation of large, transient deformations at high shear rates is the *sliding-plate viscometer.* This is shown in Figure 2.5a, and it is very simple in concept. The fluid sample is placed between two horizontal parallel plates and sheared by moving one plate relative to

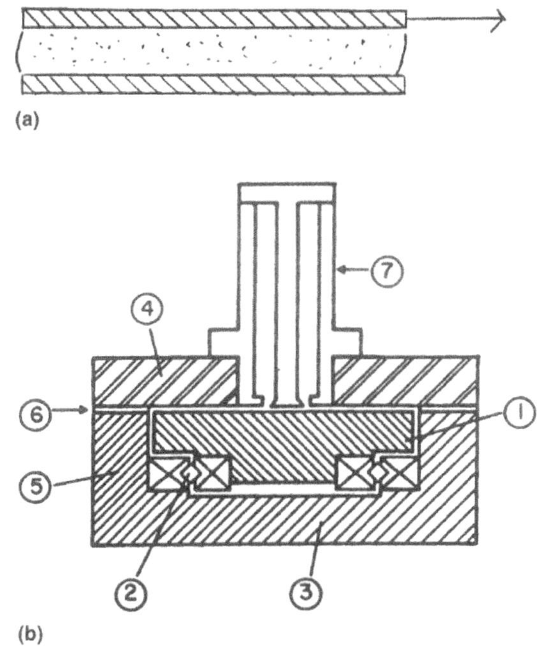

FIGURE 2.5 (a) Schematic diagram of the sliding-plate viscometer, (b) cross section of the McGill University sliding-plate rheometer, showing the moving plate, 1, linear bearing, 2, back plate, 3, stationary plate, 4, side supports, 5, shims, 6, and the shear stress transducer, 7. (From Ref. [37].)

the other. The shear stress is determined as the ratio of the measured shear force to the plate surface area; the shear rate is simply the plate velocity divided by the plate separation. The major difficulty with this experimental scheme is that stress measurements are influenced by edge effects, instrument friction, and fluctuations in plate spacing [3]. Dealy and coworkers [36, 37] overcame these problems by measuring the shear stress locally with the use of a patented shear stress transducer; capacitance proximetry was employed to measure the deflection of a cantilever beam when the shear stress acted on the free end of the cantilever. The shear stress range of the transducer was 0.01–1 MPa.

The sliding-plate rheometer developed by Giacomin et al. [37] is shown in schematic form in Figure 2.5b. The sliding plate, mounted on linear bearings, is 9 in. long and 4 in. wide. The gap between the two plates is set using shims; the shear stress is monitored at the center of the stationary plate. The entire assembly is kept inside a forced-convection oven; the maximum service temperature is 350°C. The typical sample thickness is 1 mm, and shear rates as high as 250 s^{-1} can be sustained for a total of 2 s. The total strain limitation arises due to the fixed rheometer length. Besides the measurement of steady-shear viscosity and dynamic mechanical properties, the sliding-plate viscometer can be used to study startup and shutdown of steady-shear flow, large amplitude oscillatory flow, and situations in which the shear strain increases exponentially with time. Among other things, these results are useful for examining polymer entanglement effects and the tendency of anisotropic fillers to align themselves with the flow and for quantifying any slip between the polymer sample and viscometer walls. Over the years, the capabilities of the instrument have been enhanced so that it can be used to measure the normal thrust in Large Amplitude Oscillatory Shear (LAOS) [38], polymer melt viscosity at pressures up to 70 MPa [39], and the first normal stress difference in shear at shear rates as high as 100 s^{-1} [40].

VII. FALLING BALL AND FALLING NEEDLE RHEOMETERS

Almost all the viscometers that have been described so far are fairly sophisticated instruments. If one does not have access to these rheometers, one can still obtain the flow curve of a fluid by conducting some relatively simple experiments. The essential principle of a *falling ball viscometer* can be understood based on the creeping flow of a Newtonian liquid past a stationary sphere. It is possible to determine the velocity distribution around the sphere, and this allows for the calculation of the normal and shear stress distributions. Upon integration of these quantities on the sphere surface, one gets the total drag force exerted by the flowing liquid on the sphere. This expression is very simple, and it is known as Stokes' law for drag (see, for example, [29]). The result can be employed to conduct a force balance on a sphere of radius R and density ρ_s falling at its constant terminal velocity V in an unconfined Newtonian liquid of viscosity μ and density ρ. At steady state, one finds that [29]

$$\mu = \frac{2R^2(\rho_s - \rho)g}{9V} \tag{2.22}$$

in which g is the acceleration due to gravity.

In practice, one measures the terminal velocity by dropping spheres of different diameters and densities in a cylindrical tube, and Eq. (2.22) has to be corrected for the presence of wall and end effects [41]. The form and magnitude of this correction is available [3]. This procedure works for measuring the zero-shear viscosity of non-Newtonian liquids as well since the viscosity is constant in the lower Newtonian region. The use of two or more spheres of different radii and density values can help to confirm that the measured viscosity is truly the zero-shear viscosity. In addition, several extrapolation procedures are also available to establish the zero-shear viscosity from the measured terminal velocity values [41].

To obtain the viscosity versus shear rate relationship in the shear-thinning region using falling ball data, Chhabra and coworkers [42] assumed that the flow behavior could be represented by the power-law constitutive equation. They solved for the velocity distribution in the vicinity of the sphere using numerical methods and used the results to compute the shear stress $\tau_{r\theta}$ and the corresponding shear rate at each point on the sphere surface. They then averaged the shear stress and shear rate over the entire sphere for both Newtonian and power-law fluids. In the former case, the average shear stress turned out to be

$$\frac{2gR(\rho_s - \rho)}{9}$$

while the average shear rate was $\frac{V}{R}$. The ratio of these two quantities was the Newtonian viscosity. They demonstrated the validity of this result by measuring the terminal velocity in constant viscosity liquids using a number of spheres of different radii and density values; viscosity values superposed well with values measured using a standard viscometer. For power-law fluids, they used the average shear stress and shear rate calculations to determine the ratio of the average Newtonian shear stress and shear rate to the corresponding non-Newtonian values as a function of the *power-law index*. They also showed that the power-law index could be obtained as the slope of the straight line that resulted from plotting the logarithm of the average Newtonian shear stress versus the average Newtonian shear rate but utilizing terminal velocity values obtained from experiments on polymeric liquids. Knowing the power-law index, the true shear stress versus shear rate behavior was determined from the earlier calculated relationship between the Newtonian and non-Newtonian values as a function of the power-law index. The comparison between the flow curve obtained in this manner with that using standard measurements is excellent.

The *falling needle viscometer* is a variation of the falling ball viscometer. This is shown in schematic form in Figure 2.6. Here, a hollow cylinder or needle of diameter d falls under gravity through a fluid contained in a cylinder of diameter D. We again measure the terminal velocity V, and, as expected, this depends on the needle

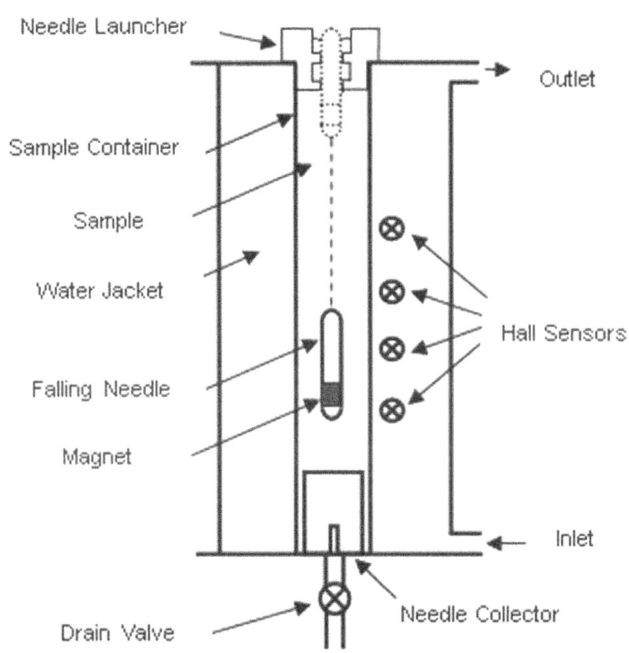

FIGURE 2.6 Schematic diagram of the falling needle viscometer. Figure courtesy of Dr. N. A. Park.

density ρ_s and the fluid density ρ. When the needle falls through a Newtonian liquid of viscosity μ, one has the following analytical result [43, 44]:

$$\mu = \left(\frac{gd^2}{8V}\right)(\rho_s - \rho)\left[-(1 + lnk)\right] \tag{2.23}$$

in which k (<0.033) is $\frac{d}{D}$, and g is the acceleration due to gravity. Once data are corrected for end effects, there is good agreement between viscosity measured using a falling needle viscometer, capillary viscometers, and rotational viscometers [43].

To obtain the flow curve for a non-Newtonian liquid, one may assume that the material behaves as a power-law liquid. Now the velocity and stress distributions near the needle surface have to be obtained numerically. According to the analysis [43], the power-law index is given by the slope of the straight line that results from plotting $\ln(\rho_s - \rho)$ versus $\ln V$, while the *consistency index*, the second model parameter, is obtained from the intercept of this line with the ordinate at $V = 1$. This information can be converted easily to viscosity as a function of shear rate as explained in [43]. Commercial versions of the instrument are manufactured by Stony Brook Scientific in Norristown, PA.

VIII. OTHER VISCOMETERS

Even though one of the stated goals of rheometry is to generate information that can guide us in optimizing polymer processing operations, which typically involve complex flow fields, material response is usually characterized in simple flow situations using viscometers of the kind described in this chapter. The use of such viscometers ensures that, as far as possible, true material behavior is determined and the measurements are independent of the technique of measurement and any assumptions regarding the constitutive behavior of the material. In some instances, these data can be related directly to polymer processability, but in other cases we have to make use of mathematical models. Rheological data then form one kind of input to the models. In industrial practice, however, one often uses indexers or testers that tend to mimic a particular processing operation, and data from these devices provide direct information about polymer processability. It is, therefore, logical to ask if these testers can also be analyzed or calibrated to yield fundamental rheological information. Here we describe two common instruments for which the answer appears to be in the affirmative.

A. Torque Rheometers

A *torque rheometer* is a batch mixer consisting of two irregularly shaped rollers or mixing elements contained in a heated mixing chamber shaped like a figure eight, as shown in Figure 2.7 [45]. The mixer is filled with a polymer sample, which is then melted. The torque needed to turn the rollers in opposite directions in the melt is recorded as a function of time at a fixed speed of rotation. Also monitored are the sample temperature and the pressure inside the mixer. Even at a fixed speed of rotation, it is obvious that the deformation rate varies from position to position within the rheometer. Also, the flow is not purely a shear flow. Consequently, data cannot

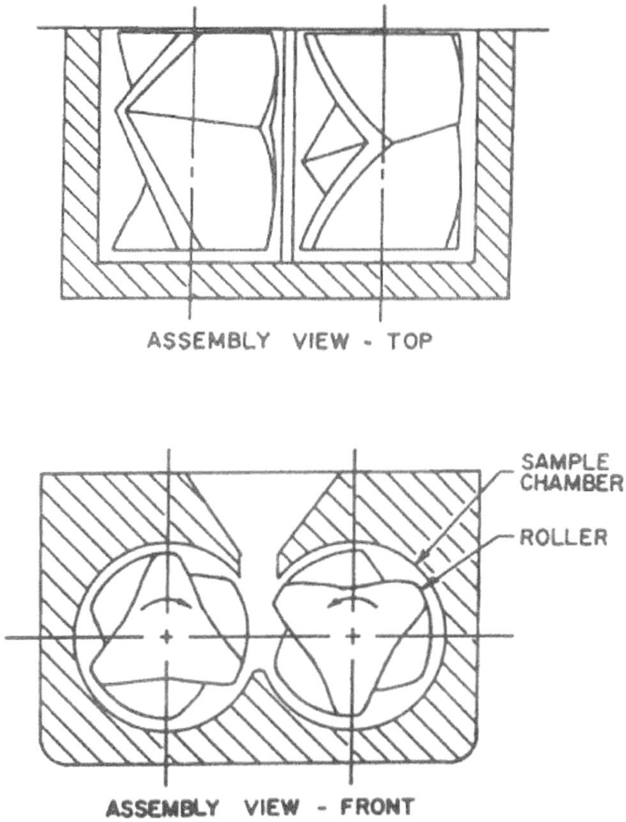

FIGURE 2.7 The Brabender torque rheometer. (From Ref. [45].)

be analyzed unless drastic assumptions are made. If it is assumed that the polymer melt is a power-law fluid and that the mixing chamber is analogous to a concentric cylinder viscometer, it can be shown that under isothermal conditions [45]:

$$M = C(n)N^n \tag{2.24}$$

in which M is the measured torque, C is a constant that depends on the mixer geometry and the power-law index n, and N is the roller revolutions per minute (rpm). Thus, torque rheometer data can reveal the power-law index n. Also, since viscosity is proportional to the torque at a fixed rpm, the temperature dependence of the viscosity can be obtained by plotting the logarithm of the torque as a function of the reciprocal of the absolute temperature.

Torque rheometers can be employed to study the processability of thermoplastics, thermosets, elastomers, and filled polymers. Data are useful for product development and quality control, in general, and in understanding the melting and degradation behavior of thermoplastics and the curing behavior of thermosets, in particular. The influence of additives can also be examined. Laboratory units have stainless steel mixing bowls of 50–150-ml capacity, and these accommodate removable mixing blades; cam, Banbury, and sigma blades are common. Measurements can be made at temperatures up to about 450°C. In the food industry, one uses a Farinograph which is an instrument that was developed in 1930. This has two Z-shaped mixing blades that rotate in opposite directions at different speeds. It is typically used to record the power needed to mix dough at a constant speed [46]. Since mixing, a common operation in the industry, establishes the rheology of the mixed mass important to subsequent processing steps and the final product quality, relating mixing power to rheology for process control is of great interest. For bread dough, analysis of the power measurement using a Gaussian model was shown to empirically relate to the dough rheology [47]. Torque rheometers are made commercially by the C.W. Brabender and Haake companies, among others. Additional details are available in the review by Chung [48].

B. Squeezing Flow Viscometer

As shown in Figure 2.8, a *squeezing flow viscometer* is an extremely simple instrument. It consists of two parallel, coaxial disks that are made to approach each other either at constant force or at constant velocity; the liquid sample is contained in the gap between the plates, and it is squeezed out of the gap in the process of testing. One measures the time dependence of the applied force F in constant velocity squeezing or the plate separation $2H$ as a function of time in constant-force squeezing. Instruments of this general nature have been around for a very long time, and the parallel-plate plastimeter [49], for example, has been used as an empirical method of getting the material properties of highly viscous polymers such as unvulcanized rubber. The measurement can be performed readily on foods using a so-called *texture analyzer* instrument, commonly available in most food labs, to elicit rheological insight. The popularity of these devices stems from the simplicity of operation and the fact that particle-and-fiber-filled, solidlike materials that might otherwise slip or fracture in conventional viscometers can be tested with ease. Squeezing flow between parallel plates also simulates the processes of compression molding and

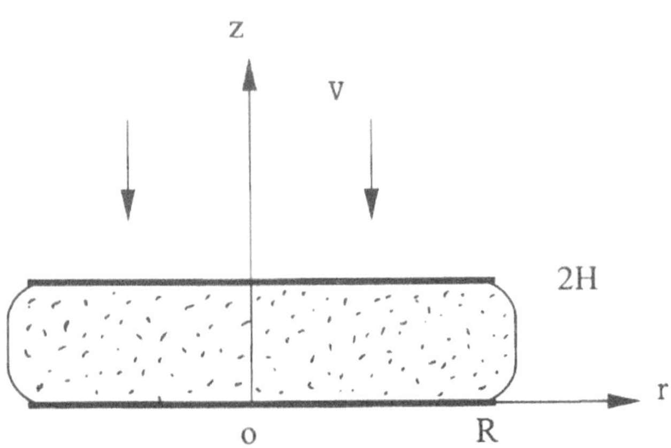

FIGURE 2.8 Schematic diagram of the squeezing flow viscometer. (From Ref. [33].)

stamping, in which we use pressure to fill a mold with a very viscous polymer charge containing fibers. The resulting flow field, however, is neither viscometric nor purely extensional. Also, the deformation is neither steady in time nor homogeneous in space. Furthermore, while the squeezing force can be predicted readily if the flow curve of the polymer is available [50], obtaining the flow curve from the measured force is possible only if we assume a form of the stress-versus-shear-rate relationship. These and other aspects of the theory have been reviewed in the literature [3, 51]. In the absence of fluid inertia, the force-versus-plate-separation relation for power-law liquids is given by the *Scott* equation [51, 52]:

$$F = \left(-\frac{dH}{dt}\right)^n \left[\frac{2n+1}{2n}\right]^n \frac{\pi K}{H^{2n+1}} \frac{R^{n+3}}{n+3} \tag{2.25}$$

in which R is the plate radius, K is the fluid consistency index, and n is the power-law index. For Newtonian liquids, n equals unity, K equals the Newtonian viscosity, and Eq. (2.25) is known as the *Stefan equation*.

The Scott equation can usually represent the response of unfilled polymeric fluids to compressive flow provided that squeezing is not rapid; for rapid squeezing (high Deborah number flows), the influence of fluid elasticity has to be taken into account as well, but this is not an easy exercise [51]. For particulate-and-fiber-filled materials, one often observes a yield stress, and the analysis leading to Eq. (2.25) can be repeated using either the Bingham plastic or *Herschel–Bulkley fluid* models [49, 53]. The resulting expressions are much more involved when compared to Eq. (2.25), but one can use the experimental data to extract the model parameters in a straightforward manner. Note that Lipscomb and Denn [54] have objected to this analysis on the grounds that Bingham-type fluids cannot contain yield surfaces in complex geometries. As a practical matter, though, these theoretical objections are not very serious, and the procedure of Covey and Stanmore [53] works quite well.

Although squeezing flow viscometers are available commercially [2], we can easily attach the parallel plates to the crosshead of a tensile testing machine and build our own instrument. We can also conduct experiments at high temperatures with the use of an isothermal oven. Since the total strain that can be imposed is small, this viscometer is best suited for making measurements at low deformation rates; under these conditions, elastic effects also are not important. In closing this section, we mention that the use of a low-viscosity lubricating liquid layer on the surface of the plates changes the mode of deformation within the more viscous material to equal biaxial extension. This lubricated squeezing flow can be utilized for measuring the *biaxial extensional viscosity* [55]. A review of the squeezing flow method can be found elsewhere [56].

IX. CONCLUDING REMARKS

Techniques used to make rheological measurements on complex fluids in shear flow are now well established. In the future, one can expect improvements in ease, accuracy, and sensitivity of measurements and the ability to make measurements over a wider range of shear rates.

The rheologist's desire is to have a single instrument that can be used to completely characterize the material rheology. However, this is not possible due to wide-ranging behavior of the rheology of complex fluids. This requires the instrument operator to not only understand the limitations and sources of errors due to the instruments and their geometries but also the nature of the sample (such as specific challenges that it may pose while loading, the sensitivity of the structure and its irreversibility, temperature dependence, heterogeneity, solvent evaporation, air entrapment, among others). Thus, the type of instrument, the geometry, the sample nature, the application purpose, not to mention experience, all need to be taken into consideration to identify the right rheological measurement for industrial use [21, 57–60].

In addition, one wishes to make measurements during phase change—melting, crystallization and gelation of polymers, for instance. In other words, we wish to couple rheology with morphology by probing changes in material structure under well-defined flow conditions. Techniques currently being experimented with for this purpose include nuclear magnetic resonance (NMR), small angle X-ray scattering (SAXS), and optical microscopy [61]. Note that these techniques are also relevant to food processing where structure necessarily changes in the process of cooking [20].

There are a number of empirical instruments in industrial use, sometimes referred to as indexers, as practical tests and often for quality control purposes [58]. These techniques generally attempt to approximate an industrially relevant flow, phenomenon, or just serve as a low-cost simple measurement tool near the production line. Examples include the melt indexer (applying the capillary rheometer principle) and the parallel-plate or "Williams" *plastometer* (applying the squeezing flow principle). There is a much larger array of empirical measurements employed in the food industry [59]; some common examples are the Bostwick *consistometer* and the Ford or Zahn cup. The flows and principles in these measurements may be classified under the term "*protorheology*" and analyzed to obtain rheological quantities [59, 62].

REFERENCES

1. J. R. Van Wazer, J. W. Lyons, K. Y. Kim, and R. E. Colwell. Viscosity and Flow Measurement. Wiley Interscience, New York, 1963.
2. J. M. Dealy. Rheometers for Molten Plastics. Van Nostrand Reinhold, New York, 1982.
3. A. A. Collyer and D. W. Clegg, eds. Rheological Measurement. 2nd ed., Chapman & Hall, London, 1998.
4. C. W. Macosko. Rheology: Principles, Measurements, and Applications. VCH/Wiley, New York, 1994.
5. R. W. Whorlow. Rheological Techniques. 2nd ed., Ellis Horwood, New York, 2002.
6. L. E. Nielsen. Polymer Rheology. Marcel Dekker, New York, 1977.
7. K. Walters. Rheometry. Chapman and Hall, London, 1975.
8. B. Rabinowitsch. Ober die viscositat und elastizitat von solen. Z. Physik. Chem. (Leipzig) 145A:1–26 (1929).
9. E. B. Bagley. End corrections in the capillary flow of polyethylene. J. Appl. Phys. 28:624–627 (1957).

10. E. B. Bagley. The separation of elastic and viscous effects in polymer flow. Trans. Soc. Rheol. 5:355–368 (1961).
11. M. Padmanabhan and M. Bhattacharya. In-line measurement of rheological properties of polymer melts. Rheol. Acta 31:31–87 (1994).
12. P. J. Carreau, D. C. R. De Kee, and R. P. Chhabra. Rheology of Polymeric Systems. 2nd ed., Hanser, Munich, 2021.
13. H. M. Laun. Capillary rheometry for polymer melts revisited. Rheol. Acta 43:509–528 (2004).
14. H. C. Brinkman. Heat effects in capillary flow I. Appl. Sci. Res. A2:120–124 (1951).
15. H. M. Laun. Pressure dependent viscosity and dissipative heating in capillary rheometry of polymer melts. Rheol. Acta 42:295–308 (2003).
16. J. Kestin, M. Sokolov, and W. Wakeham. Theory of capillary viscometers. Appl. Sci. Res. 27:241–264 (1973).
17. G. D. Galvin, J. F. Hutton, and B. Jones. Development of a high-pressure, high-shear-rate capillary viscometer. J. Non-Newtonian Fluid Mech. 8:11–28 (1981).
18. H. Munstedt. Influence of hydrostatic pressure on rheological properties of polymer melts – A review. J. Rheol. 64:751–774 (2020).
19. A. Ahuja, R. Lee, and Y. M. Joshi. Advances and challenges in the high-pressure rheology of complex fluids. Adv. Colloid Inter. Sci. 294:102472 (2021).
20. M. R. Mackley and D. G. Hassell. The multipass rheometer a review. J. Non-Newtonian Fluid Mech. 166:421–456 (2011).
21. J. Vlachopoulos and N. Polychronopoulos. Basic concepts in polymer melt rheology and their importance in processing. In: M. Kontopoulou, ed. Applied Polymer Rheology: Polymeric Fluids with Industrial Applications. Wiley, Hoboken, NJ, 2012, pp. 1–27.
22. A. V. Shenoy and D. R. Saini. Thermoplastic Melt Rheology and Processing. Marcel Dekker, New York, 1996.
23. A. V. Shenoy., S. Chattopadhyay, and V. M. Nadkarni. From melt flow index to rheogram. Rheol. Acta 22:90–101 (1983).
24. A. V. Shenoy, D. R. Saini, and V. M. Nadkarni. Rheograms for engineering thermoplastics from melt flow index. Rheol. Acta 22:209–222 (1983).
25. D. R. Saini and A. V. Shenoy. Melt rheology of some specialty polymers. J. Elastomers Plast. 17:189–217 (1985).
26. M. Padmanabhan and M. Bhattacharya. Effect of extrusion processing history on the rheology of corn meal. J. Food Eng. 18:335–349 (1993).
27. J. M. Dealy and T. O. Broadhead. Process rheometers for molten plastics: A survey of existing technology. Polym. Eng. Sci. 33:1513–1523 (1993).
28. F. Xie, P. J. Halley, and L. Averous. Rheology to understand and optimize processibility, structures and properties of starch polymeric materials. Prog. Polym. Sci. 37:595–623 (2012).
29. R. B. Bird, W. E. Stewart, and E. N. Lightfoot. Transport Phenomena. 2nd ed., Wiley, New York, 2007.
30. A. P. R. Eberle, D. G. Baird, P. Wapperom, and G. M. Velez-Garcia. Obtaining reliable transient rheological data on concentrated short fiber suspensions J. Rheol. 53:1049–1068 (2009).
31. D. M. Binding and K. Walters. Elastico-viscous squeeze films. Part 3. The torsional-balance rheometer. J. Non-Newtonian Fluid Mech. 1:277–286 (1976).
32. M. S. Carvalho, M. Padmanabhan, and C. W. Macosko. Single-point correction for parallel disks rheometry. J. Rheol. 38:1925–1936 (1994).
33. R. K. Gupta. Polymer and Composite Rheology. 2nd ed., Marcel Dekker, New York, 2000.
34. B. Maxwell and R. P. Chartoff. A polymer melt in an orthogonal rheometer. Trans. Soc. Rheol. 9:41–52 (1965).

35. H. A. Barnes. A review of the slip (wall depletion) of polymer solutions, emulsions and particle suspensions in viscometers: Its cause, character, and cure. J. Non-Newtonian Fluid Mech. 56:221–251 (1995).
36. A. J. Giacomin and J. M. Dealy. A new rheometer for molten plastics, SPE ANTEC Tech. Papers 32:711–714 (1986).
37. A. J. Giacomin, T. Samurkas, and J. M. Dealy. A novel sliding plate rheometer for molten plastics. Polym. Eng. Sci. 29:499–504 (1989).
38. J. G. Oakley and A. J. Giacomin. A sliding plate normal thrust rheometer for molten plastics. Polym. Eng. Sci. 34:580–584 (1994).
39. F. Koran and J. M. Dealy. A high pressure sliding plate rheometer for polymer melts. J. Rheol. 43:1279–1290 (1999).
40. J. Xu, S. Costeux, J. M. Dealy, and M. N. De Decker. Use of a sliding plate rheometer to measure the first normal stress difference at high shear rates. Rheol. Acta 46:815–824 (2007).
41. R. P. Chhabra and S. A. Patel. Bubbles, Drops, and Particles in Non-Newtonian Fluids. 3rd ed., CRC Press, Boca Raton, 2023.
42. P. Suri, S. A. Patel, and R. P. Chhabra. Falling ball method for determining zero shear and shear-dependent viscosity of polymeric systems: Solutions, melts, and composites. J. Vinyl Addit. Technol. 29:670–684 (2023).
43. N. A. Park and T. F. Irvine Jr. Measurement of rheological fluid properties with the falling needle viscometer. Rev. Sci. Instrum. 59:2051–2058 (1988).
44. N. A. Park and T. F. Irvine Jr. The falling needle viscometer: A new technique for viscosity measurements. Wareme Stoffubertrag. 18:201–206 (1984).
45. L. L. Blyler Jr. and J. H. Daane. An analysis of Brabender torque rheometer data. Polym. Eng. Sci. 7:178–181 (1967).
46. J. E. Bock and C. Don. The Farinograph Handbook: Advances in Technology, Science, and Applications. 4th ed., Woodhead, Cambridge, 2022.
47. F. Hugon, M. Dufour, H. Chiron, G. Della Valle, and K. Kansou. Modeling the mixing power curve to assess bread dough rheological behavior. J. Cer. Sci. 111:103667 (2023).
48. J. T. Chung. Torque rheometer technology and instrumentation. In: N. P. Cheremisinoff, ed. Encyclopedia of Fluid Mechanics. Vol. 7, Gulf, Houston, 1988, pp. 1081–1138.
49. J. R. Scott. Theory and application of the parallel-plate plastimeter. I.R.I. Trans. 7: 169–186 (1931).
50. G. Brindley, J. M. Davies, and K. Walters. Elastico-viscous squeeze films. Part I. J. Non-Newtonian Fluid Mech. 1:19–37 (1976).
51. R. K. Gupta and A. B. Metzner. Squeezing flows. Adv. Transport Processes 6:27–77 (1989).
52. P. J. Leider and R.B. Bird. Squeezing flow between parallel disks. 1. Theoretical analysis. Ind. Eng. Chem. Fundam. 13:336–341 (1974).
53. G. H. Covey and B. R. Stanmore. Use of the parallel-plate plastometer for the characterization of viscous fluids with a yield stress. J. Non-Newtonian Fluid Mech. 8: 249–260 (1981).
54. G. G. Lipscomb and M. M. Denn. Flow of Bingham fluids in complex geometries. J. Non-Newtonian Fluid Mech. 14:337–346 (1984).
55. S. Chatraei, C. W. Macosko, and H. H. Winter. Lubricated squeezing flow: A new biaxial extensional rheometer. J. Rheol. 25:433–443 (1981).
56. J. Engmann, C. Servais, and A. S. Burbidge. Squeeze flow theory and applications: A review. J. Non-Newtonian Fluid Mech. 132:1–27 (2005).
57. J. M. Dealy and K. F. Wissbrun. Melt Rheology and Its Role in Plastics Processing: Theory and Applications. Van Nostrand Reinhold, New York, 1990.
58. J. M. Dealy and P. C. Saucier. Rheology in Plastics Quality Control. Hanser, Munich, 2000.

59. M. C. Bourne. Food Texture and Viscosity: Concept and Measurement. 2nd ed., Academic, San Diego, 2002.
60. J. H. Prentice. Measurements in the Rheology of Foodstuffs. Elsevier Applied Science, London, 1984.
61. V. Rantzsch, M. B. Ozen, K. F. Ratzsch, E. Stellamanns, M. Sprung, G. Guthausen, and M. Wilhelm. Polymer crystallization studied by hyphenated Rheology techniques: RHEO-NMR, Rheo-saxs and Rheo-microscopy. Macromol. Mater. Eng. 2019, 304, 1800586.
62. M. T. Hossain and R. H. Ewoldt. Protorheology. J. Rheol. 68:113–144 (2024).

3 Shear Viscosity of Polymeric Melts

I. INTRODUCTION

At a sufficiently low temperature, every polymer acts like a hard glassy material. To process such a polymer into a fiber or a film or a molded article, it is necessary to heat it to a temperature at which it either softens (if amorphous) or melts (if crystalline) and at which it can undergo a shaping operation. Because the shear viscosity of flexible chain polymers decreases as the temperature increases, polymer processing is done at high temperatures. The particular temperature picked is based on a number of considerations. The temperature has to be high enough that the polymer flows easily, but not so high that an excessive amount of heat has to be removed when the plastic part is cooled back to room temperature, since this represents a penalty both in terms of energy consumption and processing time. In addition, the material must not thermally degrade at the chosen temperature. Besides temperature, the operating pressure and the rate of shear are also important. In injection molding, for instance, very high pressures are employed, especially in the packing stage. This results in an increase in the viscosity. During mold filling, on the other hand, the shear rate in the melt delivery system is typically about 1000 s^{-1}, while at the mold gate it may be as high as 10^5 s^{-1}. Under these conditions, the polymer may exhibit marked shear thinning, and the viscosity at high rates of shear may be several orders of magnitude smaller than the viscosity at low rates of shear. It is, therefore, necessary to measure the shear viscosity as a function of temperature, pressure, and the shear rate.

Typical curves for the steady-shear viscosity of a molten polymer as a function of temperature and shear rate are shown in Figure 3.1 on logarithmic coordinates. These are often called flow curves. Notice that at high temperatures, the melt has a large but constant viscosity at low rates of shear; this region is known as the lower Newtonian region, and it is characterized by a temperature-dependent zero-shear-rate viscosity η_0. However, over much of the usual accessible shear rate range, particularly at the lower temperatures, the viscosity decreases linearly with shear rate in a double logarithmic graph. In this linear range, the so-called power-law equation holds, in which the shear stress and the viscosity are, respectively, given by:

$$\tau = K\dot{\gamma}^n \tag{3.1}$$

$$\eta_a = K\dot{\gamma}^{(n-1)} \tag{3.2}$$

where the constant K is called the consistency index and the constant n is known as the power-law index. The subscript a on η indicates that the viscosity given by Eq. (3.2) is the apparent viscosity rather than the true viscosity. For Newtonian liquids,

DOI: 10.1201/9781003013501-3

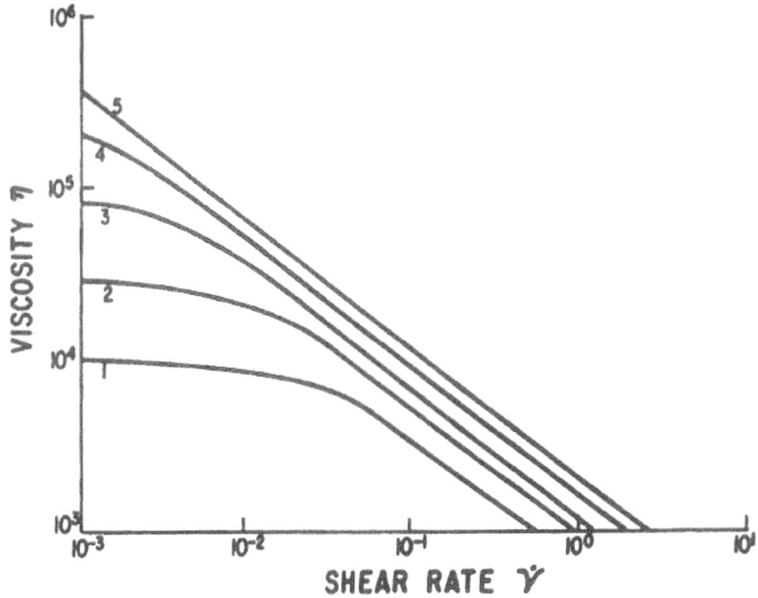

FIGURE 3.1 Typical log viscosity–log shear rate curves at five different temperatures. Curve 1 is for the highest temperature; curve 5 is for the lowest temperature. For a typical polymer, the temperature difference between each curve is approximately 10°C. (From Ref. [2].)

n equals unity and K is the same as the viscosity; polymer melts, however, are shear thinning or pseudoplastic, and the value of n is less than 1. The power law generally does not hold accurately over the entire range of shear rates, but in the range where the equation is valid, the slope of the straight lines shown in Figure 3.1 is $(n - 1)$. Equation (3.2) obviously cannot represent the viscosity over the entire range of shear rates, because it predicts a meaningless increase in the viscosity as the shear rate is decreased. A popular alternative model that does not have this shortcoming is the *Carreau equation* [1]:

$$\frac{\eta_a}{\eta_0} = \left[1 + (\lambda_c \dot{\gamma})^2\right]^{(n-1)/2} \tag{3.3}$$

which has three constants: η_0, λ_c, and n. Where η_0 accommodates a limiting shear viscosity at low shear rates, the parameter λ_c determines the point of onset of shear thinning; as shown in Figure 3.1, the onset of shear thinning moves to larger values of the shear rate as the temperature of measurement is increased.

Equation (3.3) was generalized by Yasuda et al. by adding a fourth constant "a", and the resulting equation is known as the *Carreau-Yasuda model* [3, 4], such that for $a = 2$, the Carreau model is recovered:

$$\frac{\eta_a - \eta_\infty}{\eta_0 - \eta_\infty} = \left[1 + (\lambda_c \dot{\gamma})^a\right]^{(n-1)/a} \tag{3.4}$$

Another equation that is often used to fit shear viscosity data over the entire range of shear rates is the *Cross equation* [4, 5]:

$$\eta_a = \frac{\eta_0}{1+(\lambda_c\dot{\gamma})^{1-n}} \tag{3.5}$$

where η_0 and n have the same significance as in Eq. (3.3), and λ_c is a fitting parameter. It is easy to show that when $\lambda_c\dot{\gamma} = 1$, $\eta_a = \eta_0/2$.

Polymer melts are obviously non-Newtonian, in the sense that their viscosity depends on the rate of shear. They are also non-Newtonian in the sense that they are viscoelastic; i.e., they have some properties of elastic solids. After inception of shear flow at a constant shear rate, for example, the shear stress takes some time to reach a steady state and can even overshoot the steady-state value. Similarly, upon cessation of steady shearing, the shear stress takes a finite amount of time, known as the *relaxation time*, to go to zero. As a consequence, the shear viscosity can depend on time even when the deformation rate is kept constant. We defer an examination of viscoelastic effects to later chapters; here we focus on the steady-shear viscosity alone.

For a given polymer, as explained earlier, the viscosity depends on processing variables such as the temperature, pressure, and the shear rate. However, the viscosity value changes as the chemical nature of the polymer changes. Thus, the shear viscosity of polyethylene differs from that of nylon, even when the values of the three processing variables are kept unchanged. Additionally, for a given polymer type, the viscosity is a function of the molecular weight, the molecular weight distribution and the presence of any chain branching. For food polymer melts, such as thermoplastic starch, the nature and amount of plasticizer used and the thermomechanical treatment, since they are typically processed under high shear and temperature conditions (using extruders, for example), are equally important variables [6]. In the sections that follow, we systematically examine the influence of each of these variables. We present typical data, show how these data can be combined to give viscosity master curves, propose equations for data representation, and provide physical explanations for the recorded observations. Both model and commercial polymers are considered. As might be expected, data are most abundant for high-volume polymers such as polyethylene, polypropylene, and polystyrene. Note that while the polyolefins are crystalline, polystyrene is amorphous, and the molecular weight of all three polymers, when manufactured commercially, is high enough that their rheological behavior is quite non-Newtonian. Also note that these three polymers are normally synthesized using chain growth (addition) polymerization, and they are reasonably stable at processing temperatures. As opposed to this, other important polymers like polyesters and nylons are relatively Newtonian due to their lower molecular weight. Also, they are made by the process of step growth (condensation) polymerization; at processing temperatures, they are subject to hydrolysis or postcondensation, depending on the extent to which they are dried before being processed. Since most polymers tend to oxidize when kept at elevated temperatures for extended periods of time, it is normal to use an inert atmosphere of nitrogen or argon gas when making rheological measurements. Commercial polymers also contain an additive package that may include an antioxidant. Food polymers, on the other hand, are natural, available from a wide range of sources, and of many types. The effect of the food polymer melt structure on rheology is difficult to study directly using rheological methods. However, the principles outlined in the following sections have been successfully applied to food polymeric systems [6, 7].

II. ZERO-SHEAR VISCOSITY

A major difference between the rheological behavior of polymer melts and low-molecular-weight materials is that the shear viscosity of polymeric fluids depends on the rate of shear. This was seen in Figure 3.1, and we will pursue this aspect in detail in the next section. Here we examine the zero-shear viscosity or the limiting behavior at low shear rates. Note that it is sometimes necessary to extrapolate data to the zero-deformation rate, because the use of a viscometer at hand may not always allow for the unequivocal determination of the zero-shear viscosity. Alternately, dynamic mechanical data may be employed to compute the zero-shear viscosity, since [8]

$$\eta_0 = \lim_{\omega \to 0} \frac{G''(\omega)}{\omega} \tag{3.6}$$

where G'' is the loss modulus defined by Eq. (1.8). The variables that influence the zero-shear viscosity of synthetic polymers are polymer molecular weight, molecular weight distribution, chain branching, temperature, and pressure.

A. Influence of Molecular Weight and Molecular Weight Distribution

It is possible to study the influence of molecular weight on the melt viscosity of linear polymers, since the technique of anionic polymerization, which yields materials of narrow molecular weight distribution, allows for the synthesis of essentially monodisperse polymers such as polystyrene and polybutadiene. Of course, a polydisperse sample can be fractionated, but this is a tedious procedure that results in only small quantities of material. When isothermal, zero-shear-viscosity data are measured as a function of polymer molecular weight, it is found that below a critical molecular weight M_c that is independent of temperature, the zero-shear viscosity is roughly proportional to the molecular weight M_w. That is,

$$\eta_0 = K_1 M_w \quad M_w < M_c \tag{3.7}$$

At molecular weights above M_c, the zero-shear viscosity depends upon M_w to a power equal to about 3.4 or 3.5:

$$\eta_0 = K_2 M_w^{3.4} \quad M_w > M_c \tag{3.8}$$

This behavior is illustrated in Figure 3.2 for polystyrene melts at 217°C [9] and is representative of polystyrene in general [10]. It is also found to hold for other polymers [11]. To emphasize the generality of Eqs. (3.7) and (3.8), Graessley and

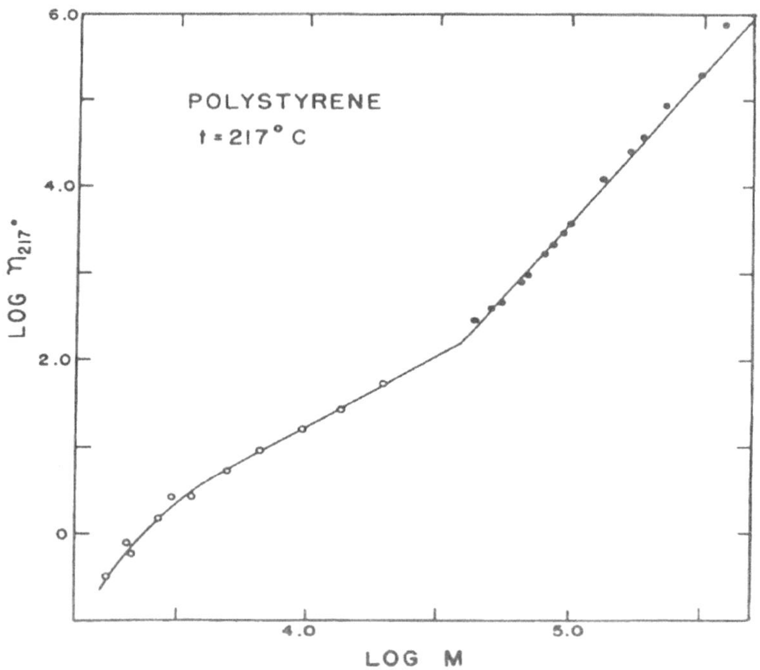

FIGURE 3.2 Double logarithmic plot of zero-shear viscosity as a function of the polymer molecular weight for polystyrene fractions. (From Ref. [9]. Copyright © 1954 by John Wiley & Sons, Inc. Reprinted by permission of John Wiley & Sons, Inc.)

coworkers reanalyzed available data on polyethylene and showed that the relationship of melt viscosity versus molecular weight was not influenced by the ability of the polymer to crystallize [12]. The truth of this conclusion was further demonstrated by presenting data on (amorphous) poly butadiene [13] and (crystalline) hydrogenated polybutadiene [14].

The transition between Eqs. (3.7) and (3.8) is usually not sharp. However, data can generally be represented by two straight lines that intersect at $M = M_c$; in Figure 3.2, M_c equals approximately 38,000. For most polymers, M_c lies between about 10,000 and 40,000 and increases with increasing chain stiffness. Some typical values are given in Table 3.1. The change in slope of the plot of viscosity versus molecular weight shown in Figure 3.2 is explained by saying that when it exceeds a critical molecular weight, polymer chains begin to get entangled with each other and the entanglement density increases with increasing molecular weight [15, 16]. As sketched in Figure 3.3, *entanglements* may be visualized as temporary crosslinks whose effect is to prevent chain crossability and to increase fluid friction vastly. This effect of entanglements is analogous to the fact that it requires more force to pull a strand from a ball of yarn when the strands are entangled and knotted than when they are not entangled. Thus, flow becomes much more difficult because forces applied to one polymer chain become transmitted to and distributed among many other chains. Explanations based on polymer chain entanglements have now been superseded by

TABLE 3.1
Entanglement Molecular Weight M_c

Polymer	M_c
Linear polyethylene (PE)	4000
Polyisobutylene	17,000
Polyvinyl acetate	29,200
Polystyrene	38,000
Polydimethyl siloxane	35,200
Polymethyl methacrylate	10,400
Caprolactam polymer (linear)	19,200
Caprolactam polymer (tetra-branched)	22,000
Caprolactam polymer (octa-branched)	31,100

Source: From Ref. [2].

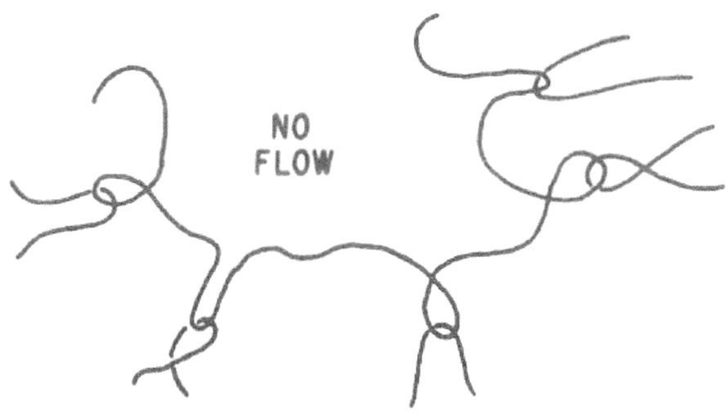

FIGURE 3.3 Schematic diagram of entanglement formation in quiescent polymer melts. (From Ref. [2].)

detailed analyses based on tube models (see [17], for a short review). However, the concept of entanglements is a useful aid to thinking just as combinations of springs and dashpots are useful for describing polymer viscoelasticity.

With the foregoing picture of chain entanglements in mind, it is logical to inquire whether the critical chain length or the number of monomer units needed for entanglement formation is the same for most polymers. It is found that although there is some variability in the value of the critical chain length Z_c amongst polymers, the variation in Z_c is much smaller than the variation in M_w [9]; Z_c corresponds to about 300–700 main chain atoms [18]. Furthermore, since the molecular weight is linearly related to the polymer chain length, M_w in Eqs. (3.7) and (3.8) can be replaced by the chain length. Indeed, when data on a number of polymers are plotted as the logarithm of the zero-shear viscosity versus the logarithm of the chain length, the change in slope occurs at essentially the same value of Z [19]. The significance of this observation is that the zero-shear viscosity of any polymer can be predicted as a function of molecular weight if one is able to compute the temperature-dependent critical viscosity corresponding to M_c or Z_c in Figure 3.2. One way of doing this is through the group contribution method of van Krevelen and te Nijenhuis [20].

In order to explain the observed relationship between the zero-shear viscosity and the molecular weight on a fundamental level, one has to appeal to molecular models. In one of the earliest theories, due to Debye and Bueche [21–23], the shear viscosity was calculated by determining the amount of energy dissipated due to fluid friction in a steady laminar shearing flow at a constant shear rate. The analysis is very straightforward and leads to the expected proportionality between the zero-shear viscosity and the molecular weight; the constant of proportionality involves the monomer molecular weight and unknown quantities, such as polymer chain dimensions and the friction coefficient per monomer unit. In the presence of entanglements, it is assumed that the motion of one molecule is resisted by the neighboring molecules that are dragged along with it but move at a lower velocity. The neighboring molecules, in turn, snag other molecules, which also move but at a still slower velocity. When all the resistances are added, the result is Eq. (3.8). Even though this is exactly the anticipated result, the notion of one chain dragging several other chains leads to an incorrect dependence of the mean relaxation time on the molecular weight [24]. Correcting this internal inconsistency requires the use of a more sophisticated model. At the present time, the accepted picture of the motion of a polymer molecule in an entangled melt is one in which the surrounding molecules constrain the molecule of interest to lie within a tube so that it can move readily only along its own length. This slithering motion is known as *reptation* [24, 25], and it is the central feature of a molecular theory developed by Doi and Edwards [26]. The progression from the concept of entanglements to the idea of reptation has been chronicled by Marrucci [27]. The theory is fairly involved, and working through the details requires a certain facility with mathematics. While an explicit expression can be derived for the zero-shear viscosity, the dependence on the molecular weight turns out to be to the third power rather than to the expected 3.4 power [28]. Nonetheless, the reptation model provides a remarkably consistent interrelationship between various viscoelastic functions, which has been confirmed with observations on linear polymers [29]. Tube models have evolved over the years [30], and new developments are reported on a regular basis.

For polydisperse, linear polymers, we expect the zero-shear viscosity to depend on the molecular weight distribution or, at the very least, on one or more averages of the molecular weight distribution. Measurements on binary blends of narrow-molecular-weight-distribution polystyrene [31, 32] and on samples of wide-molecular-weight-distribution polyethylene [33] at a variety of temperatures have demonstrated that the weight average molecular weight $\bar{M}_w$ is the important molecular weight. Note that $\bar{M}_w$ is the molecular weight averaged with respect to the weight fraction. It is found that Eqs. (3.7) and (3.8) remain valid when the weight average molecular weight is used in place of M_w. This unique dependence of the zero-shear viscosity on $\bar{M}_w$ has also been observed for silicone polymers [34], polypropylene [35], and nylon 6 [36], and it leads to experimentally verifiable blending rules for the viscosity [34]. For linear polyethylenes, though, the exponent on $\bar{M}_w$ is found to be 3.6 [37]. Also, there are some published data that suggest that when the molecular weight distribution is very wide, as in commercial polymers, the melt viscosity may not be strictly dependent upon $\bar{M}_w$ [38, 39]. Rather, the logarithm of the zero-shear viscosity may be proportional to an average molecular weight in between $\bar{M}_w$ and the next higher moment of the distribution curve beyond $\bar{M}_w$. This implies that the high-molecular-weight "tails" on the distribution curve can be especially important in affecting the zero-shear viscosity and other rheological properties.

B. Influence of Chain Branching

Polyethylene is the highest volume polymer globally, and when this is commercially made in a high-pressure tubular reactor using free radical initiators, one obtains a polydisperse structure containing random long branches. This is known as low-density polyethylene (LDPE); the low density in the solid phase is due to the inability of the irregular branched molecules to pack efficiently into a crystal lattice. When ethylene is polymerized employing Ziegler–Natta catalysts, whether in a slurry reactor or a fluidized bed reactor, the result is a linear polymer but one that has a broad molecular weight distribution [40]. This is known as high-density polyethylene (HDPE). An even broader molecular weight distribution, having a polydispersity index (PDI) of 12–24, can be attained with the help of a "Phillips" catalyst [40]. The *polydispersity index* is defined as $\bar{M}_w/\bar{M}_n$, where $\bar{M}_n$ is the number-average molecular weight which is the molecular weight averaged with respect to the mole fraction. Linear LDPE (LLDPE) has short branches, and it is made like HDPE but by reacting ethylene with 1-butene, 1-hexene, or 1-octene. More recently, the development of (soluble) metallocene catalysts that have an active center has allowed for the synthesis of long-chain molecules with virtually no branching [40]. Here, the PDI is 2 [41]. For branched molecules also, the use of metallocene catalysts yields a lower polydispersity, and for LLDPE, a much more homogeneous comonomer distribution with near-perfect regularity [41] is obtained. In addition, one can synthesize polyethylene with low contents of long-chain branching [42]. This great variety of structures influences the rheological as well as the mechanical properties of the polymer. For branched polymers, the branches can be long or short, and these, in turn, can have secondary branching. The branches can be randomly spaced along the backbone chain, or several branches can originate from a single point to give a star-shaped molecule. Some of these possibilities are sketched in Figure 3.4. While the comb structure could be considered to be representative of LLDPE made by copolymerizing ethylene with a monomer such as butene-1, random branching is more characteristic of traditional LDPE. For controlled rheological studies, however, the specially synthesized star-branched molecule has proven popular [43]. An interesting variation is a dendritic macromolecule: a treelike structure with branches emanating from a multifunctional core and from each unit of monomer [44].

Short branches generally do not affect the viscosity of a molten polymer very much. On the other hand, long branches can have a very large effect [46]. Branches that are long but that are still shorter than those required for entanglements decrease the zero-shear viscosity when compared to a linear polymer of the same molecular weight [47–49]. Such branching reduces the viscosity because branched molecules are more compact than linear molecules. It is found that the zero-shear viscosity can

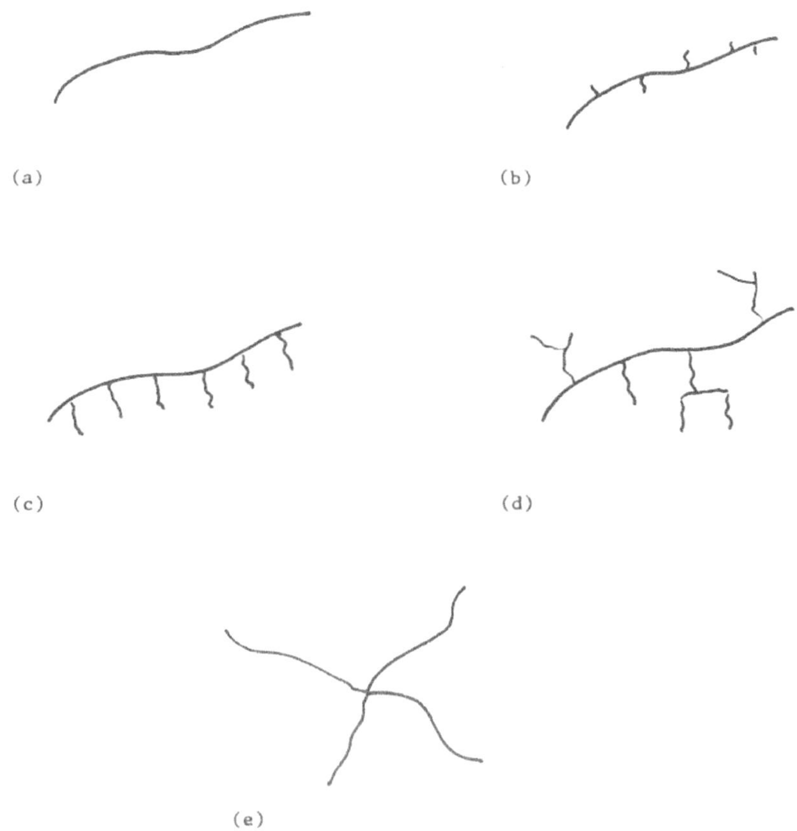

FIGURE 3.4 Chain branching in polymers: (a) linear molecule; (b) short branches; (c) comb structure; (d) long (random) branches; (e) star-branched molecule. (From Ref. [45].)

be considered the product of two terms, one the monomeric friction coefficient and the other a structure factor that is proportional to the number of interaction points [18, 43]. Since individual molecules in a molten polymer assume the shape of a random coil, one has the situation depicted in Figure 3.5 [50], and it is clear that there is a lesser degree of molecular overlap, resulting in a lower viscosity due to fewer

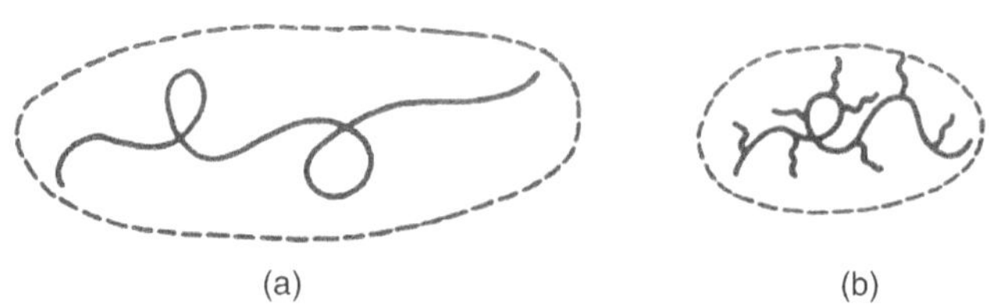

FIGURE 3.5 Oversimplified view of how the hydrodynamic volume varies for (a) a linear polymer and (b) a branched polymer of the same molecular weight. (Reproduced with permission from Ref. [50]; copyright © 1981, Division of Chemical Education, Inc.)

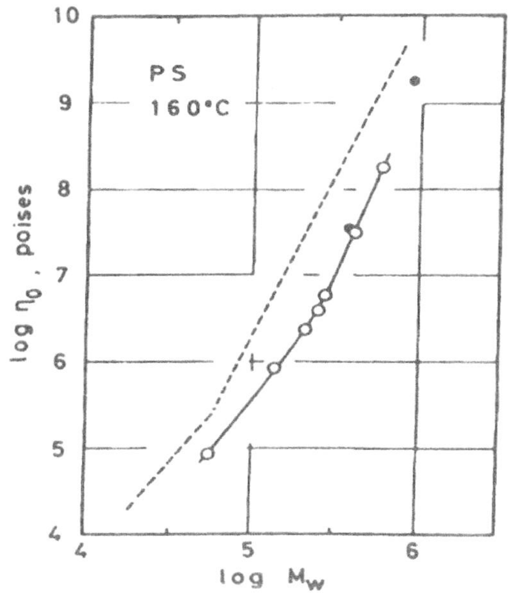

FIGURE 3.6 Molecular weight dependence of zero-shear viscosity for star-branched polystyrenes. (Reprinted with permission from Ref. [55]. Copyright 1971 American Chemical Society.)

intermolecular interactions between branched macromolecules. However, if the branches are so long that they can participate in entanglements, the branched polymer may have a viscosity at low rates of shear that is significantly greater than that of a linear polymer of the same molecular weight [51, 52]. This conclusion appears to be true even when the number of branches is below the detection limit of techniques, such as ^{13}C-NMR [53].

Thus, the molecular weight of branches is more important than their number. It is these branching differences that cause the zero-shear viscosity of high-density polyethylene to be higher than that of LLDPE but lower than that of LDPE [54]. A logical consequence of these arguments is that the zero-shear viscosity of a low-molecular-weight branched polymer must be lower than that of a linear polymer of comparable molecular weight, but it should increase much more rapidly with molecular weight as the length of the branches is increased. This is seen to be true in Figure 3.6 for four-branched star polystyrenes; the slope of the plot for the branched polymer is 4.5 rather than the 3.4 for the linear polystyrene.

C. Influence of Temperature

As mentioned earlier, a polymer, whether amorphous or crystalline, is a hard, brittle, glassy solid at low temperatures. At sufficiently high temperatures, on the other hand, an uncrosslinked polymer turns into a viscous liquid. The transition between these two states is not sharp; there is an intermediate rubbery stage in which the material behaves like an elastomer or a soft elastic solid. The temperature at which a hard, glassy polymer becomes a rubbery material is called the *glass transition temperature*, T_g. In physical terms, the glass transition temperature is interpreted as the temperature below which only the vibrational motion of atoms or the short cooperative motion of a small collection of atoms can take place. Immediately above T_g, however, the polymer backbone can flex and large recoverable deformations are possible, because the molecule can uncoil but not flow. The rubbery region extends over a significant temperature range because flow is hampered by the presence of polymer chain entanglements that act like temporary crosslinks. In view of all this, it is not surprising that the viscosity of most polymers changes greatly with temperature.

For ease of flow, a polymer molecule must have enough thermal energy to make it mobile or able to get away from its neighbors, and, in addition, there must be enough space around it to allow it to move past other polymer molecules [23]. Below T_g, the latter condition is not met, and the polymer is a solid. Above T_g the magnitude of the shear viscosity is dependent on the availability of *free volume*, which, on a microscopic level, is that part of the sample volume not actually occupied by the polymer molecules. The free volume is zero at 0 K, and it increases with increasing temperature. If one assumes that the ratio of the volume of a polymer chain segment to the free volume associated with that segment is the same for all amorphous polymers at the glass transition temperature, it is possible to show that [20, 23]

$$\log \frac{\eta_0(T)}{\eta_0\left(T_g\right)} = \frac{-17.44\left(T - T_g\right)}{51.6 + \left(T - T_g\right)} \tag{3.9}$$

Equation (3.9) is known as the WLF equation, after Williams, Landel, and Ferry, who first proposed it [56, 57]. The viscosity at T_g is often about 10^{12} Pa-s. Equation (3.9) is found to be approximately applicable to all polymers in a temperature range between T_g and $T_g + 100\ K$.

At temperatures far above the glass transition temperature or the melting point for crystalline polymers, there is ample free volume available, and the temperature dependence of the zero-shear viscosity is determined by energy barriers to motion. Under these circumstances, the viscosity follows the *Andrade* or *Arrhenius* equation to a good approximation:

$$\eta_0 = Ke^{E/RT} \tag{3.10}$$

In this equation, K is a constant characteristic of the polymer and its molecular weight, E is the activation energy for the flow process, R is the universal gas constant, and T is the temperature in degrees Kelvin. The activation energy generally is in the range between 2.09×10^4 and 2.09×10^5 J/mole. Table 3.2 lists some typical values of the energy of activation for several polymers. The value of 16.7 kJ/mol for dimethyl silicone polymers is the smallest known; this low value is the result of the great flexibility of the silicone polymer chain. The energy of activation for flow increases as the size of the side groups increases and as the chain becomes more rigid [58, 59]. For starch-based melts (starch, corn meal, and wheat starch), the activation energy reported range from 1.2×10^4 to 1.5×10^5 J/mol [6].

TABLE 3.2
Energy of Activation for Flow of Polymers

Polymer	Energy of Activation *E* kcal/mol	kJ/mol
Demethyl silicone	4	16.7
Polyethylene (high density)	6.3–7.0	26.3–29.2
Polyethylene (low density)	11.7	48.8
Polypropylene	9.0–10.0	37.5–41.7
Polybutadiene (cis)	4.7–8	19.6–33.3
Polyisobutylene	12.0–15.0	50–62.5
Polyethylene terephthalate	19	79.2
Polystyrene	25	104.2
Poly (α-methyl styrene)	32	133.3
Polycarbonate	26–30	108.3–125
Poly (1-butene)	11.9	49.6
Polyvinyl butyral	26	108.3
SAN (styrene acrylonitrile copolymer)	25–30	104.2–125
ABS (20% rubber) (acrylonitrile-butadiene-styrene copoymer)	26	108.3
ABS (30% rubber)	24	100
ABS (40% rubber)	21	87.5

Source: From Ref. [2].

To illustrate the enormous temperature dependence of the viscosity, let us use the Andrade equation with $E = 20{,}000$ cal/mole. On raising the temperature from 300 to 310 K, the viscosity decreases by a factor of about 2.96! Figure 3.7 shows how the viscosity changes with temperature according to the WLF equation. The Andrade equation would give a nearly straight line on the same type of plot. If the viscosity and its temperature coefficient are the same for the WLF and Andrade equations at 100 K above T_g, the viscosity according to the Andrade equation would be given approximately by the dashed line in Figure 3.7. As already mentioned, the Andrade equation is satisfactory if the temperature is greater than $(T_g + 100)$; at lower temperatures, the WLF equation should be used with an adjustable empirical value for $\eta_0(T_g)$.

For liquids obeying the Andrade equation, the energy of activation E is nearly independent of the temperature. However, for the WLF equation, the apparent energy of activation or the temperature coefficient of viscosity not only depends upon temperature but also upon the glass transition temperature:

$$E = \frac{Rd(\log \eta_0)}{d(1/T)} = \frac{4.12 \times 10^3 T^2}{\left(51.6 + T - T_g\right)^2} \tag{3.11}$$

Table 3.3 gives E for various values of T_g and $(T - T_g)$. The energy of activation becomes very large as the temperature approaches T_g, especially if T_g is large.

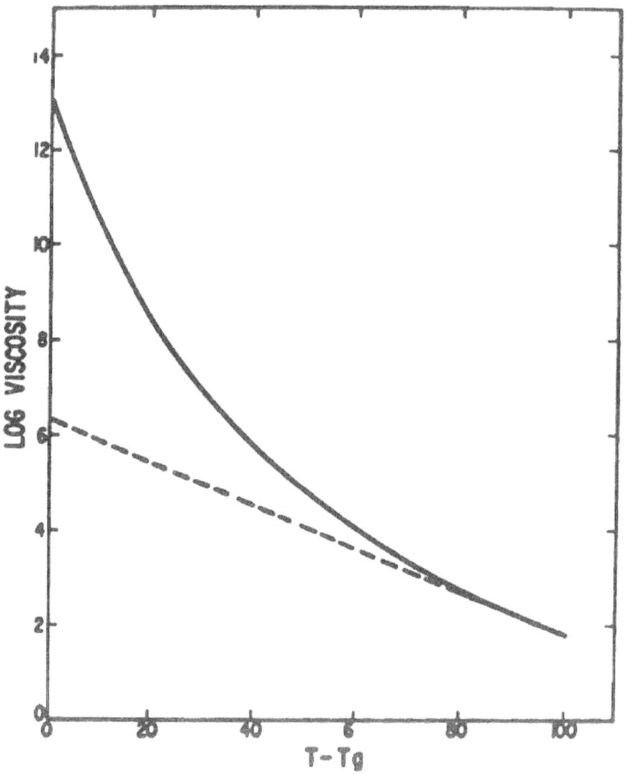

FIGURE 3.7 Viscosity as a function of temperature according to the WLF equation. The dotted line gives the approximate viscosity dependence according to the Arrhenius equation when matched to the WLF value at (T_g + 100 K). (From Ref. [2].)

TABLE 3.3
WLF Energy of Activation for Viscous Flow

$T - T_g$	T_g = 200 K	T_g = 250 K	T_g = 300 K	T_g = 350 K
0	61.9	96.7	139.3	189.6
2	58.5	91.1	130.8	177.7
5	54.1	83.6	119.6	162.1
10	47.9	73.4	104.3	140.7
20	38.9	58.6	82.3	110.0
30	32.7	48.5	67.4	89.4
50	25.0	35.9	48.9	63.9
80	18.7	25.9	34.4	44.0
100	16.1	22.0	28.7	36.3

Source: From Ref. [2].

If the energy of activation for use in the Andrade equation of a particular polymer is not known, it can be estimated by assigning values to each of the groups making up the monomeric unit and adding up the values for all the groups [20, 60]. The energy of activation E is given by

$$E = \left[\frac{\sum X_i}{M_0}\right]^3 \tag{3.12}$$

The molecular weight of the monomeric unit is M_0, and X_i is an empirical value assigned to group i. Values of X_i have been tabulated by van Krevelen and te Nijenhuis [20] and van Krevelen and Hoftyzer [60].

Thus far, we have considered the influence of temperature on the steady-shear viscosity when this quantity is measured at constant temperature. Commercial polymer processing operations, however, can involve extremely large rates of change of temperature following a fluid element. If, under rapidly varying temperature conditions, the instantaneous value of material functions such as the shear viscosity is still given by Eq. (3.9) evaluated at the instantaneous value of the temperature, we say that the polymer melt is *thermorheologically simple* [61]. More generally, though, even if the deformation rate is kept constant while the temperature is raised or lowered, stress transients of fairly large magnitude can arise [62, 63]. A partial explanation of this phenomenon is provided by the observation that the value of the glass transition temperature that appears in Eq. (3.9) is itself influenced by the rapid variation in temperature; for polystyrene melts, a T_g value that is 12–16°C higher than the equilibrium value has been measured for cooling rates in the range of 0.5–4.5°C/sec [62].

D. Influence of Pressure

Even though liquids are normally considered incompressible, this is not strictly true. During polymer processing operations such as extrusion and injection molding, the applied pressure can reach a thousand atmospheres [64]; such a large hydrostatic pressure results in a decrease in the free volume in the melt, and, according to several theories [65–67], should lead to a significant increase in the viscosity, especially above but not far from the polymer glass transition temperature. In this sense, an increase in pressure can be considered equivalent to a decrease in the temperature [68]. Indeed, the combination of high pressure and low temperature tends to induce crystallization in polymers such as polyethylene. It has been suggested that Eq. (3.10) for the temperature dependence of viscosity should be modified to [66]:

$$\eta_0 = K \exp\left(\frac{E}{RT} + \frac{CV_0}{V_f}\right) \tag{3.13}$$

in which V_0 is approximately the occupied volume at the glass transition temperature and V_f is the increase in volume due to thermal expansion above T_g. Thus,

$$V_f = \alpha V_0 \left(T - T_g\right) \tag{3.14}$$

where α is the coefficient of thermal expansion. The value of the constant C in Eq. (3.13) is generally between 0.5 and 1.0.

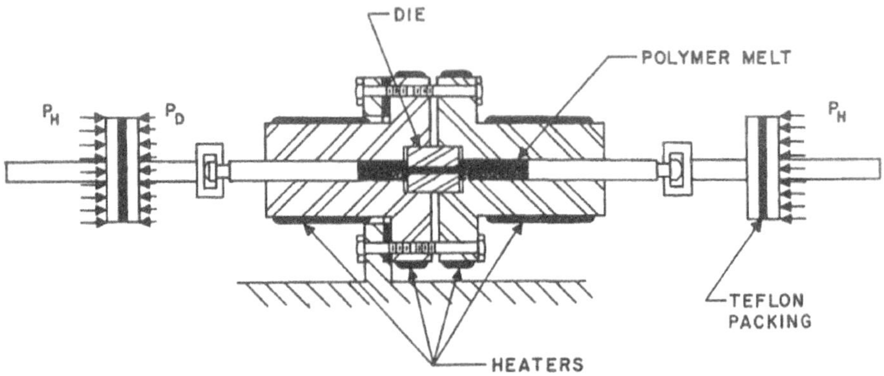

FIGURE 3.8 Schematic diagram of the hydrostatic pressure viscometer. (From Ref. [69].)

Due to the difficulty in making measurements, data about the effect of pressure on the viscosity of polymer melts are not plentiful. It is, however, generally accepted that the viscosity increases with pressure p according to [4]:

$$\eta_0 = \eta_{ref} e^{\alpha p} \tag{3.15}$$

in which α is of the order of 2×10^{-8} Pa^{-1}.

Westover has described a double-piston apparatus for measuring viscosity as a function of pressure up to 25,000 psi, and this is shown in schematic form in Figure 3.8 [69]. The hydrostatic pressure exerted on the polymer sample is determined by the pressure of the oil at the extreme ends of the hydraulic cylinders. The application of a known differential pressure across the die causes flow to take place through the die. The resulting volumetric flow rate is measured by knowing the volume swept out by the ram at the entrance side of the die. Data are analyzed according to the procedure described in Section II of Chapter 2, Capillary Viscometers. Results obtained on a LDPE sample at three different temperatures and two die sizes are displayed in Figure 3.9 [69]. It is seen that increasing pressure has a dramatic effect on the measured viscosity, particularly as the temperature is lowered. In other studies of the influence of pressure on viscosity, McGowan and Cogswell [70] employed a pressurized concentric cylinder viscometer, while Kamal and Nyun [71] proposed a method of analyzing capillary viscometer data to extract the pressure dependence of the viscosity. A capillary instrument to make viscosity measurements at high pressures was manufactured commercially earlier by Polymer Laboratories. On the theoretical side, Utracki [72] has shown that the zero-shear viscosity data of McGowan and Cogswell [70] on polyethylene, polystyrene, and PMMA can be satisfactorily explained within the framework of the free volume theory. Other developments have been described by Chakravorty et al. [73]. Later, Park and Dealy used a high-pressure sliding-plate rheometer to measure the viscosity versus shear rate behavior of a HDPE melt at 180°C and at pressures ranging from atmospheric to 69 MPa [74]; the low-shear-rate viscosity more than doubled over this pressure range.

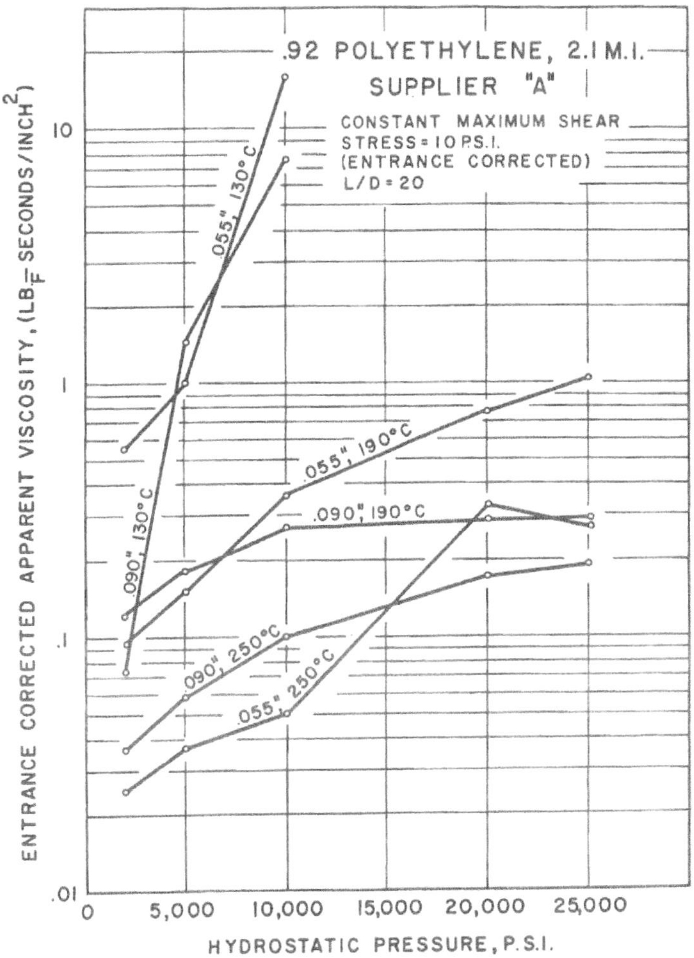

FIGURE 3.9 Viscosity versus hydrostatic pressure at three temperatures and two die sizes. (From Ref. [69].)

III. SHEAR RATE-DEPENDENT VISCOSITY

An outstanding characteristic of polymer melts is their shear-thinning behavior whereby the viscosity decreases as the rate of shear increases. This non-Newtonian behavior is of tremendous practical importance in the processing and fabrication of plastics and elastomers. First, the decreased viscosity makes the molten polymer easier to process or squirt through small channels, such as in the filling of a mold. In other words, the additional energy required to operate a large injection molding machine or extruder is reduced due to this phenomenon. Second, the decrease in viscosity is associated with the development of elasticity, represented by N_1, in the melt. This elasticity produces such phenomena as *die swell*, or the "puff-up" of extruded strands. Molecular orientation in molded objects also is closely related to polymer elasticity.

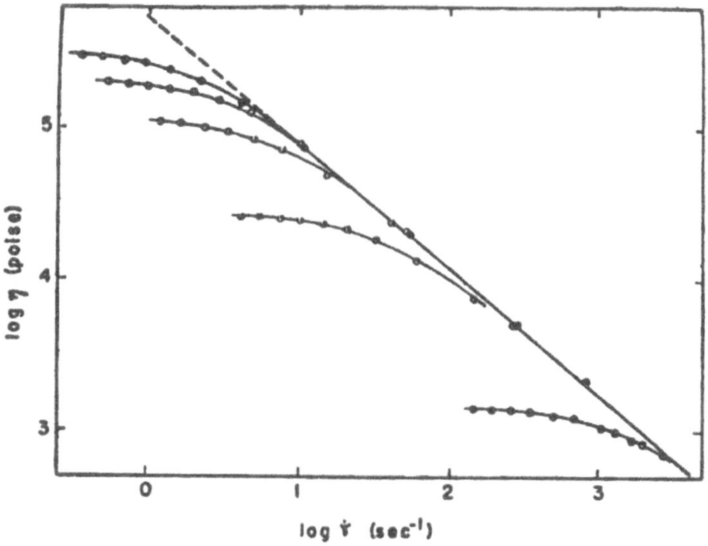

FIGURE 3.10 Viscosity as a function of shear rate for polystyrene melts of different molecular weights at 183°C. From left to right, the curves are for molecular weights of 242,000, 217,000, 179,000, 117,000, and 48,500. (From Ref. [75].)

Typical curves for the viscosity as a function of shear rate for a number of different temperatures are shown in Figure 3.1. Actual data, again as a function of shear rate but at a constant temperature, are shown in Figure 3.10 for monodisperse polystyrene melts of different molecular weights [75]. Figure 3.10 reveals that the zero-shear viscosity increases with molecular weight, as expected. Since all the molecular weights are larger than the critical value for entanglement formation, shear thinning is observed in each case. Because low-molecular-weight polymers have fewer entanglements than high-molecular-weight ones, it is not surprising that the point of onset of shear thinning moves to smaller values of the shear rate with increasing molecular weight. Furthermore, it is found that all the curves merge into a single line at high shear rates, implying a constant value of the power-law index.

The data displayed in Figures 3.1 and 3.10 can almost always be reduced to a single *master curve* if the ratio of the viscosity to the zero-shear viscosity (i.e., the relative viscosity) is plotted in terms of an appropriate dependent variable. Such a treatment is also suggested by Eqs. (3.3)–(3.5); it merely remains to define λ or λ_c. In the simplest case, a unique curve is obtained if the logarithm of the relative viscosity is plotted against the logarithm of the product of the shear rate with the zero-shear viscosity, as shown in Figure 3.11; the effect of temperature largely is automatically compensated for by using the relative viscosity instead of the viscosity itself and by replacing the shear rate with the shear stress in the Newtonian region. Kraus and Gruver [76] found this to be true for random copolymers of butadiene and styrene. Since the ordinate as well as the abscissa are scaled with the zero-shear viscosity, whose behavior with both molecular weight and temperature is well understood, obtaining the shear rate-dependent viscosity with the help of such a master curve is

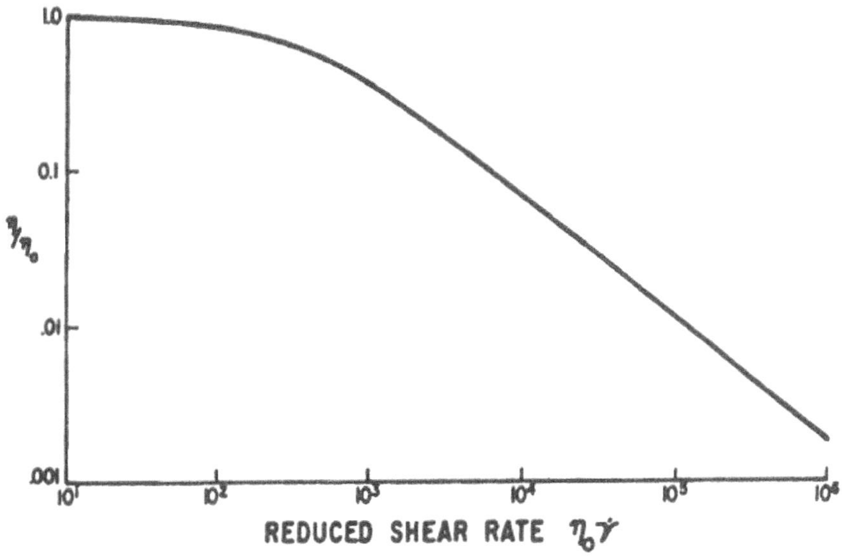

FIGURE 3.11 The master curve produced from the data of Figure 3.1 by vertical and horizontal shifting of the curves. (From Ref. [2].)

a straightforward matter. In other cases, better fit of the data is obtained if the abscissa scale is log $\left(\dot{\gamma}\eta_0 M_w^\alpha/\rho T\right)$; the constant α is generally near unity [77], but it may vary somewhat. The temperature T is in Kelvin, and ρ is the density of the polymer at each temperature [4, 57]. Such a master curve is shown in Figure 3.12 employing the data of Figure 3.10 [75].

There are several theories and resulting equations that often are used to describe the shear rate dependence of the viscosity. These include the theories of *Bueche* [23] and of Graessley [16, 78, 79], with the latter theory being the more successful one. In the *Graessley theory*, the viscosity is calculated by dividing the energy dissipated per unit time per unit volume by the square of the shear rate, due to steady shearing at a constant shear rate. The energy dissipation is itself computed by determining the fluid drag at each entanglement point and adding together the contributions from all the polymer–polymer entanglements in unit volume of fluid. A lowering in viscosity with increasing shear rate is attributed to shear-induced reduction in the entanglement density. The final result for the relative viscosity is

$$\frac{\eta}{\eta_0} = \frac{2}{\pi}\left[\cos^{-1}\alpha + \frac{\alpha\left(1-\alpha^2\right)}{\left(1+\alpha^2\right)^2}\right] \tag{3.16}$$

in which $\alpha = \left(\eta/\eta_0\right)\left(1/2\dot{\gamma}\theta_0\right)$, where θ_0 is a time constant for the formation of entanglements at low shear rates. As with Eq. (3.3) earlier, Eq. (3.16) suggests that a unique curve should be obtained if the relative viscosity is plotted against $\dot{\gamma}\theta_0$. This is generally found to be the case.

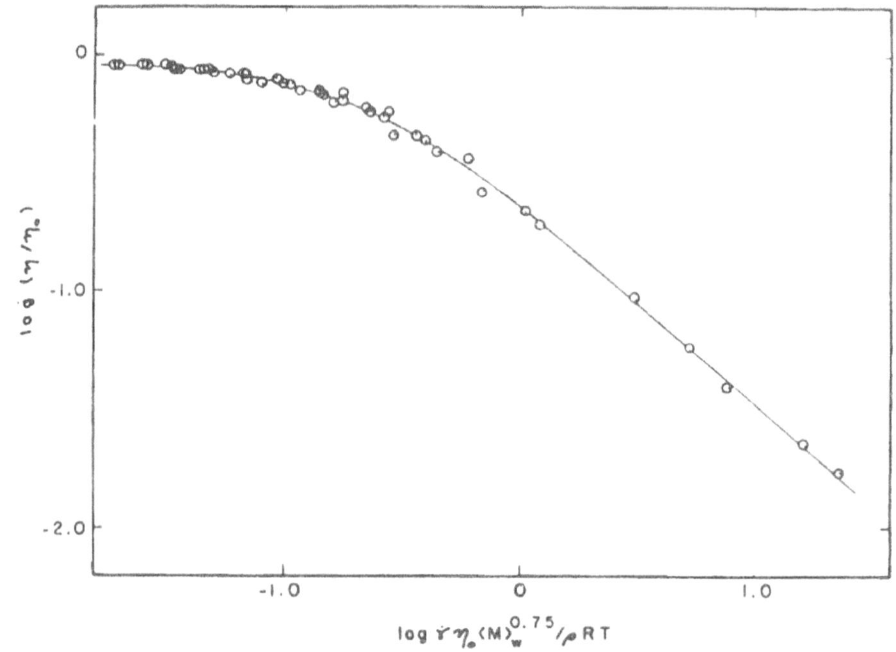

FIGURE 3.12 Logarithmic plot of reduced viscosity against reduced shear rate, using the data of Figure 3.10. (From Ref. [75].)

As far as tube models are concerned, the essential idea is that a molecule in a polymer melt finds itself surrounded by neighboring molecules that rob it of its freedom of movement. The molecule of interest, or the test chain, finds itself situated inside a tube which is oriented in a particular direction. As a result, the molecule can diffuse only along the tube axis and not perpendicular to it. Upon deformation of the polymer, the tube orientation can change, and, within the tube, the test chain can stretch or contract. To obtain the stress as a function of time, one has to determine the time evolution of the average tube orientation and the time evolution of the average chain stretch. Over the years, a very large number of models have been proposed to carry out this calculation [80]. Once the orientation tensor and the chain stretch scalar are available, the stress can be calculated at any time instant and it is possible to predict a shear-thinning viscosity [81]. Tube models, however, are still a subject of active research.

A. Influence of Polydispersity and Molecular Architecture

If $\bar{M}_w$ is the same, a distribution of molecular weights does not affect the zero-shear viscosity, but it does affect the value of the shear rate at which non-Newtonian behavior becomes apparent. It is observed that the polymer with the broader distribution exhibits non-Newtonian flow at a lower rate of shear than a polymer having

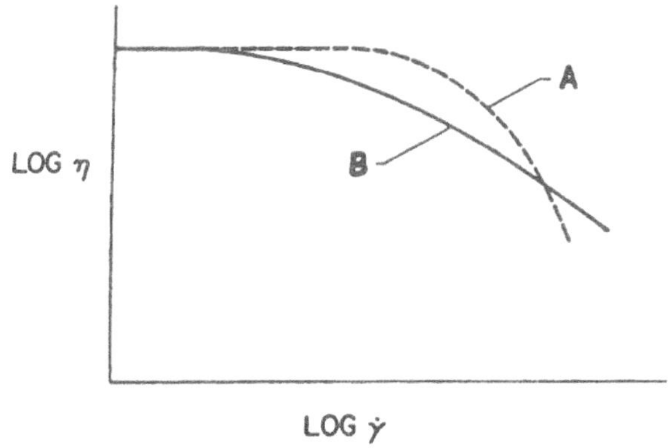

FIGURE 3.13 Plot of generalized viscosity versus shear rate for (A) a narrow-molecular-weight-distribution polymer and (B) a branch-molecular-weight-distribution polymer. The average molecular weight in the two cases is assumed to be the same. (Reproduced with permission from Ref. [50]; copyright © 1981, Division of Chemical Education, Inc.)

a narrower distribution of molecular weights [77, 82]; as mentioned, the zero-shear viscosity will be identical in these two cases. This is illustrated in Figure 3.13 [50], where it is seen that shear thinning is generally more pronounced in the polymer possessing the narrower molecular weight distribution. Often, the curves of the two samples approach each other at high shear rates [46]. The width of the distribution curve can be expressed by the PDI, which increases as the distribution broadens. The shift in non-Newtonian behavior to lower shear rates as the distribution broadens has important practical implications. Fractions or polymers with a very narrow distribution appear to have a higher viscosity than the same polymer with a normal or broad distribution under conventional molding conditions, where the rate of shear is high. Thus, polymers with a normal or broad distribution of molecular weights are generally easier to extrude or mold than a polymer with a narrow distribution.

The effect of chain branching on the shear rate-dependent viscosity is similar to the effect of chain branching on the zero-shear-rate viscosity. Thus, at high rates of shear, branched polymers in nearly all cases have lower viscosity than linear ones of the same molecular weight [46]. Although, increasing the number of branches tends to decrease the viscosity, the molecular weight of the branches seems to be more important than their number [83].

In view of the foregoing, the manufacturer of polymers has the means of controlling the zero-shear-rate viscosity and the shear rate dependence of viscosity by variations in molecular weight, molecular weight distribution, and length of branches. In particular, the lack of shear thinning of polyethylene made by the use of metallocene catalysts can be overcome by the introduction of long-chain branches. These structural factors control the rheology, which, in turn, affects the processing or fabrication and the mechanical characteristics of finished objects [49].

B. Master Curves for Temperature Dependence

Several empirical methods have been developed for predicting the temperature dependence and shear rate dependence of the shear viscosity of a polymer melt from a limited amount of experimental data. These have been motivated by the observation that isothermal curves for the variation of shear stress as a function of the shear rate at different temperatures can be made to superimpose by means of a horizontal shift, provided logarithmic coordinates are used. Such data on LDPE are shown in Figure 3.14 [84]; note that the shear rates have been calculated with the assumption of Newtonian flow behavior, although this does not influence the superposition procedure. The curve at one convenient temperature T_R can be chosen as the reference curve. In Figure 3.14 the curve at 200°C was selected as the reference curve. All the other curves can then be superimposed on the reference curve by shifting them horizontally along the shear rate axis. Curves for temperatures above the reference temperature are moved to the left; curves for lower temperatures are shifted to the right. The resulting master curve is shown in Figure 3.15. It is important to know how much each curve had to be shifted to superimpose upon the master curve and how this shifting depended upon the temperature. With this information, the procedure can be reversed by using the master curve to calculate the flow behavior at any arbitrary temperature. Note that the temperature dependence of the extent of shifting can actually be related to the temperature dependence of the zero-shear viscosity with the help of the following argument.

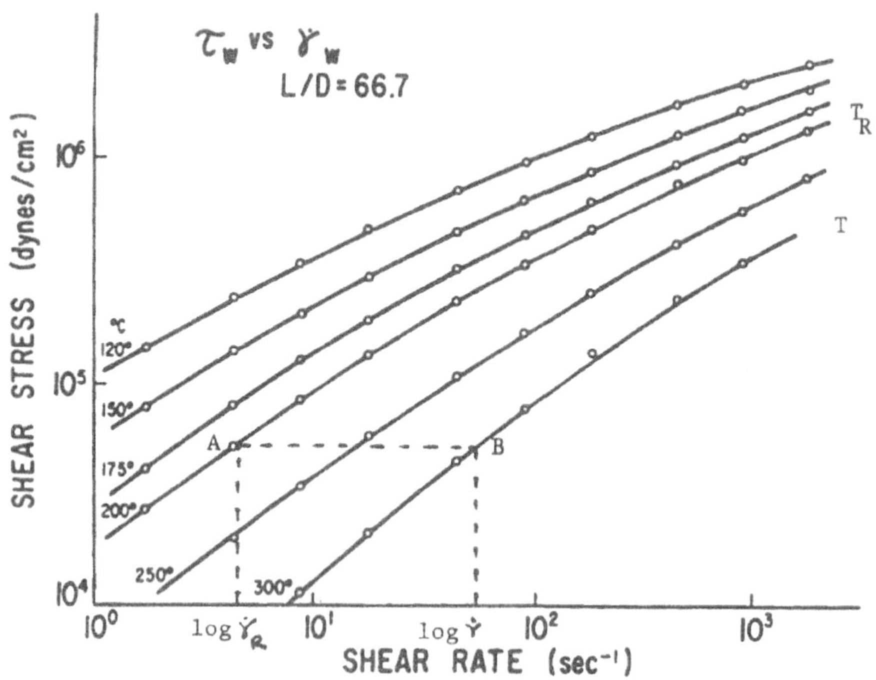

FIGURE 3.14 Flow curves at various temperatures for a low-density polyethylene. (From Ref. [84].)

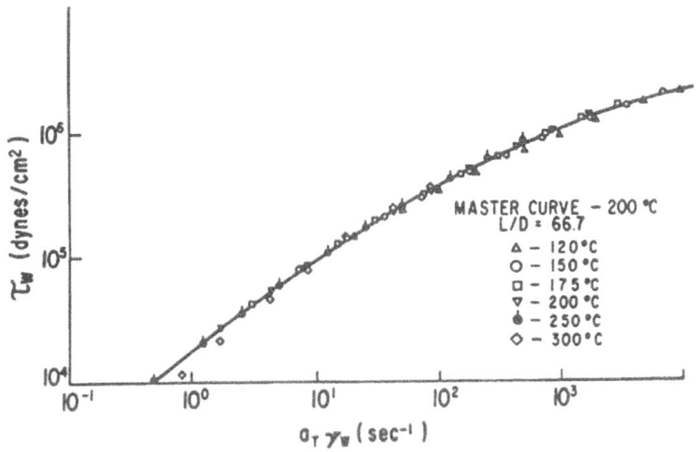

FIGURE 3.15 Master curve produced from the data of Figure 3.14. (Modified from Ref. [84].)

Let us pick two points A and B having the same ordinate value, as shown in Figure 3.14. Let the abscissa of point A be $\log \dot{\gamma}_R$ and that of point B be $\log \dot{\gamma}$, and let the distance between them, measured from the reference curve to the curve of interest, be $\log a_T$. Clearly then

$$\log \dot{\gamma}_R - \log \dot{\gamma} = \log a_T \tag{3.17}$$

And $\log a_T < 0$ when $T > T_R$. If both the points are in the linear (zero-shear) region,

$$\log\left[\frac{\eta_0(T)}{\eta_0(T_R)}\right] = \log\left[\frac{\tau(T,\dot{\gamma})/\dot{\gamma}}{\tau(T_R,\dot{\gamma}_R)/\dot{\gamma}_R}\right] = \log\left(\frac{\dot{\gamma}_R}{\dot{\gamma}}\right) = \log a_T \tag{3.18}$$

If we now evaluate Eq. (3.9) at $T = T_R$ and subtract the result from Eq. (3.9) as originally written, then

$$\log\left[\frac{\eta_0(T)}{\eta_0(T_R)}\right] = \log a_T = \frac{-C_1(T-T_R)}{C_2+(T-T_R)} \tag{3.19}$$

in which $C_2 = 51.6 + T_R - T_g$ and $C_1 = 17.44 \times 51.6/C_2$. Since the *temperature shift factor* a_T is the same in the linear as well as the nonlinear region, the second equality in Eq. (3.19) is true in general. Even though the use of universal constants appearing in Eq. (3.9) gives reasonable results, better agreement with experimental data is achieved if C_1 and C_2 are specially chosen for each polymer [57]. At high temperatures, of course, Eq. (3.10) needs to be used in place of Eq. (3.9) in the derivation leading to Eq. (3.19).

Temperature-shear rate superposition is a very powerful technique, since for many polymers the shift factors are independent of polymer molecular weight [84]. Furthermore, the same shift factors work for other viscoelastic functions as well.

C. Influence of Plasticizers

Every polymer exhibits a glass-to-rubber transition at a temperature known as the glass transition temperature, T_g. Below T_g, an amorphous polymer behaves like a hard, brittle, glassy elastic, solid, whereas above T_g, the same polymer becomes soft and rubbery. As explained earlier, the reason for this dramatic change is that the motion of polymer chains is frozen in the glassy region, and chain segments do not have "elbow room" to move past each other. Above T_g, polymer chain segments become mobile due to a progressive increase in free volume with increasing temperature. The polymer free volume is defined as the volume not occupied by the molecules themselves. Amorphous polymers begin to flow easily at sufficiently high temperatures, and many commercial polymers are processed at temperatures that are about 100°C above T_g. The use temperature of most polymers, however, is well below T_g or the crystalline melting point, as appropriate; under these conditions, the polymers are rigid. There are many instances, though, where flexibility is desirable, such as with polyvinyl chloride (PVC). This polymer finds application in tubing and hoses, shower curtains and upholstery, for example, and it is necessary to reduce stiffness since the T_g of PVC ranges from 70 to 100°C. This is accomplished by reducing T_g to a value below room temperature. The ability to do this is rooted in the observation that the ratio of the volume of the polymer chain segment that moves to the free volume associated with that segment is a constant at T_g [85, 86]. Since free volume decreases with decreasing temperature, any additive that increases free volume will necessarily result in a reduction in T_g. These additives are known as external *plasticizers*, and they "pry" polymer chains apart by forming secondary bonds with the polymer and by reducing polymer–polymer interactions; for best results, intermolecular forces between plasticizer and polymer should be of the same order of magnitude as those between polymer and polymer [87]. In addition, the plasticizer should have low volatility and be miscible with the polymer. Miscibility or compatibility can be judged in a variety of ways with the help of thermodynamics, including with the use of Hildebrand or Hansen solubility parameters and the Flory–Huggins interaction parameter [86]. The lowering of T_g is generally proportional to the plasticizer concentration, but there are many instances where polymers contain plasticizing liquids at a mass concentration of only a percent or so. Often the glass transition temperature of a liquid is about two-thirds of its melting point in degrees Kelvin. The mixture glass transition temperature lies between the T_g values of the components and can be estimated by one of the following mixture rules [88]:

$$T_g = T_{gp}\phi_p + T_{gL}\phi_L \tag{3.20}$$

$$1/T_g = W_p/T_{gp} + W_L/T_{gL} \tag{3.21}$$

where T_{gp} is the glass transition temperature of the polymer and T_{gL} is the corresponding value for the plasticizer. ϕ_p and ϕ_L are the volume fractions of the polymer and plasticizer, respectively, while W_p and W_L are the weight fractions of the polymer and plasticizer, respectively. Note that liquids also have glass transition temperatures, generally well below 0°C and in some cases far below (–100°C). Knowing the change in the glass transition temperature, we can estimate the change in viscosity with the help of Eq. (3.9).

It is estimated that almost 90% of all plasticizers produced in the world are used in the production of flexible PVC [89], and nearly all of these are esters. Indeed, di-2-ethylhexylphthalate, diisononylphthalate, and diisodecylphthalate account for nearly 75% of all the plasticizers used with PVC [90], with a typical loading level being 15 wt%. Another polymer that has gained commercial importance is polylactic acid due to the fact that it is considered to be both sustainable and industrially compostable. It is, however, a brittle polymer with low ductility. Consequently, plasticization of this polymer to improve toughness has been the subject of much research. Plasticizers that have been tried include polyethylene glycol, triethyl citrate, and acetylated tributyl citrate (ATBC). An issue with these and other plasticizers is that they tend to migrate from the interior to the surface, and their effect is lost. One approach to address this has been with the help of reactive plasticizers [91].

It is recognized that food molecules also exhibit non-equilibrium glassy solid and rubbery liquid states, and the transition from one to the other can be influenced by plasticizers. As a result, plasticizers are used extensively in the food industry and for the same reasons as in the plastics industry. Now, however, water is the most common plasticizer, especially for hydrophilic molecules, and the study of structure–property relations in foods in the presence of water has come to be known as "food polymer science". The similarities between synthetic polymers like polyethylene and natural polymers such as starch and gluten have been documented by Slade and Levine in their very extensive review on the topic, and these have been used to explain the properties of food materials during processing and storage [92]. The effectiveness of water as a plasticizer can be judged from the fact that the T_g of wheat gluten reduces from greater than 160°C at 1 wt% water to about 15°C at 16 wt% water [92].

Since fossil resources are non-renewable, efforts have been made to develop plastics based on agricultural products, and, in this regard, starch has been widely studied. Starch sourced from various grains and tubers consists of two polymers, viz., amylose and amylopectin, packed into granules which may range in size from 1 to 100 μm, depending on the source. Amylose is a linear amorphous polymer with a molecular weight ranging from 10^5 to 10^6 and amylopectin is a highly branched semicrystalline polymer with molecular weight ranging from 10^7 to 10^9; amylose content can vary from less than 1% to more than 70% [6]. Since starch decomposes without melting at high temperatures, its granules are heated in water to plasticize the amorphous regions followed by melting the crystalline regions, leading to the swelling of the granules and eventually the leaching out of the amylose polymers. This entire process is known as *gelatinization*. In the presence of shear and limited water, a homogeneous viscous liquid, called *thermoplastic starch* (TPS), is formed. Since water is a plasticizer, a brittle solid is obtained on drying TPS. Consequently, a non-volatile plasticizer, such as glycerol, may also be added to starch to maintain desirable properties in the solid state.

The most common rheological property that is measured for TPS characterization is the steady-shear viscosity. Since TPS is created in-situ during processing, such as extrusion, measuring its viscosity using rotational rheometers described in

Chapter 2 is not possible. On-line viscometers or capillary rheometers are the preferred methods for accurate measurements. One solution is to employ a slit die viscometer attached to a flood-fed single screw extruder [93]. The screw rotation speed is kept constant, and this keeps the specific mechanical energy unaltered. The shear rate at the die wall is changed by the use of a valve that varies the flow rate through the slit die. While this method yields consistent results, it does require a large sample size. One could use a multi-pass rheometer, but this is a specialized instrument, and it may not be readily available. In terms of results, TPS viscosity is found to follow power-law behavior, with viscosity decreasing with an increase in temperature. The major effect of temperature is on the consistency index; the power-law index remains relatively unaffected by changes in temperature [6].

IV. CONCLUDING REMARKS

This chapter has examined the influence of processing variables such as temperature, pressure, and shear rate and the influence of structural variables such as molecular weight, molecular weight distribution, and chain branching on the steady-shear viscosity of melts of flexible chain polymers. Typical data were presented and explained, and it was shown how viscosity master curves could be generated. This process has become standard for synthetic polymers such as polyethylene. In the case of food polymers such as starch, the transformation from the granular state to a polymeric melt requires the combination of a plasticizer (such as water), heat, and shear. Processing equipment such as an extruder is used and the hence the viscosity is measured in-situ using on-line slit die viscometers.

We emphasize that we have not explicitly considered transient effects in this chapter. These are important in food processing due to structural changes, and they are also important in polymer processing operations where the time to reach steady state may be comparable to or even longer than the time scale of the processing operation being analyzed. Finally, we point out that in the processing of semi-crystalline polymers, the rheological behavior, in a temperature range between the melting point and the glass transition temperature, is influenced by the presence of crystallinity in the melt. There is a great paucity of data under these conditions, and it is common to account for the presence of crystals in an empirical manner [94].

REFERENCES

1. R. B. Bird, R. C. Armstrong, and O. Hassager. Dynamics of Polymeric Liquids. Vol. 1. 2nd ed. Wiley, New York, 1987, p. 171.
2. L. E. Nielsen. Polymer Rheology. Marcel Dekker, New York, 1977.
3. K. Yasuda, R. C. Armstrong, and R. E. Cohen. Shear flow properties of concentrated solutions of linear and star branched polystyrenes. Rheol. Acta 20:163–178 (1981).
4. J. Vlachopoulos, and N. Polychronopoulos. Basic concepts in polymer melt rheology and their importance in processing. In: M. Kontopoulou, ed. Applied Polymer Rheology: Polymeric Fluids with Industrial Applications. Wiley, Hoboken, NJ, 2012, 1–27.
5. M. M. Cross. Polymer rheology: Influence of molecular weight and polydispersity. J. Appl. Polym. Sci. 13:765–774 (1969).

6. F. Xie, P. J. Halley, and L. Averous. Rheology to understand and optimize processibility, structures and properties of starch polymeric materials. Prog. Polym. Sci. 37:595–623 (2012).
7. L. Slade and H. Levine. The "Food Polymer Science" approach to the practice of industrial R&D, leading to patent estates based on fundamental starch science and technology. Crit. Rev. Food Sci. Nutr., 58:972–992 (2018).
8. K. Walters. Rheometry. Chapman and Hall, London, 1975, p. 30.
9. T. G. Fox and P. J. Flory. The glass temperature and related properties of polystyrene. Influence of molecular weight. J. Polym. Sci. 14:315–319 (1954).
10. A. Casale, R. S. Porter, and J. F. Johnson. Dependence of flow properties of polystyrene on molecular weight, temperature, and shear. J. Macromol. Sci.-Rev. Macromol. Chem. C5:387–408 (1971).
11. T. G. Fox and V. R. Allen. Dependence of the zero shear melt viscosity and the related friction coefficient and critical chain length on measurable characteristics of chain polymers. J. Chem. Phys. 41:344–352 (1964).
12. V. R. Raju, G. G. Smith, G. Marin, J. R. Knox, and W. W. Graessley. Properties of amorphous and crystallizable hydrocarbon polymers. I Melt rheology of fractions of linear polyethylene. J. Polym. Sci.: Polym. Phys. Ed. 17:1183–1195 (1979).
13. W. E. Rochefort, G. G. Smith, H. Rachapudy, V. R. Raju, and W. W. Graessley. Properties of amorphous and crystallizable hydrocarbon polymers. II. Rheology of linear and star-branched polybutadiene. J. Polym. Sci.: Polym. Phys. Ed. 17:1197–1210 (1979).
14. H. Rachapudy, G. G. Smith, V. R. Raju, and W. W. Graessley. Properties of amorphous and crystallizable hydrocarbon polymers. III. Studies of the hydrogenation of polybutadiene. J. Polym. Sci.: Polym. Phys. Ed. 17:1211–1222 (1979).
15. R. S. Porter and J. F. Johnson. The entanglement concept in polymer systems. Chem. Rev. 66:1–27 (1966).
16. W. W. Graessley. The entanglement concept in polymer rheology. Adv. Polym. Sci. 16: 1–179 (1974).
17. E. Narimissa and M. H. Wagner. Review on tube model based constitutive equations for polydisperse linear and long-chain branched polymer melts. J. Rheol. 63:361–375 (2019).
18. G. C. Berry and T. G. Fox. The viscosity of polymers and their concentrated solutions. Adv. Polym. Sci. 5:261–357 (1968).
19. T. G. Fox, S. Grateh, and S. Loshaek. Viscosity relationships for polymers in bulk and in concentrated solution. In: F. R. Eirich, ed., Rheology. Vol. 1. Academic Press, New York, 1956, pp. 431–493.
20. D. W. van Krevelen and K. te Nijenhuis. Properties of Polymers: Their Correlation with Chemical Structure; Their Numerical Estimation and Prediction from Additive Group Contributions. 4th ed., Elsevier Scientific, Amsterdam, 2009.
21. P. Debye. The intrinsic viscosity of polymer solutions. J. Chem. Phys. 14:636–639 (1946).
22. F. Bueche. Viscosity, self-diffusion and allied effects in solid polymers. J. Chem. Phys. 20:1959–1964 (1952).
23. F. Bueche. Physical Properties of Polymers. Wiley, New York, 1962.
24. P. G. de Gennes. Scaling Concepts in Polymer Physics. Cornell University Press, Ithaca, NY, 1979.
25. P. G. de Gennes. Reptation of a polymer chain in the presence of fixed obstacles. J. Chem. Phys. 55:572–579 (1971).
26. M. Doi and S. F. Edwards. The Theory of Polymer Dynamics. Clarendon Press, Oxford, 1986.
27. G. Marrucci. Molecular modeling of flows of concentrated polymers. Adv. Transp. Process. 5:1–36 (1987).
28. W. W. Graessley. Some phenomenological consequences of the Doi-Edwards theory of viscoelasticity. J. Polym. Sci.: Polym. Phys. Ed. 18:27–34 (1980).

29. W. W. Graessley. Entangled, linear, branched and network polymer systems—Molecular theories. Adv. Polym. Sci. 47:67–117 (1982).
30. J. M. Dealy and R. G. Larson. Structure and Rheology of Molten Polymers. Hanser, Munich, 2006.
31. J. F. Rudd. The effect of molecular weight distribution on the rheological properties of polystyrene. J. Polym. Sci. 44:459–474 (1960).
32. V. R. Allen and T. G. Fox. Viscosity-molecular weight dependence for short chain polystyrene. J. Chem. Phys. 41:337–343 (1964).
33. R. A. Mendelson, W. A. Bowles, and F. L. Finger. Effect of molecular structure on polyethylene melt rheology. 1. Shear-dependent viscosity. J. Polym. Sci.: Part A-2 8:127–141 (1970).
34. E. M. Friedman and R. S. Porter. Polymer viscosity-molecular weight distribution correlations via blending: For high molecular weight poly(dimethyl siloxanes) and for polystyrenes. Trans. Soc. Rheol. 19:493–508 (1975).
35. W. Minoshima, J. L. White, and J. E. Spruiell. Experimental investigation of the influence of molecular weight distribution on the rheological properties of polypropylene melts. Polym. Eng. Sci. 20:1166–1176 (1980).
36. H. M. Laun. Prediction of elastic strains of polymer melts in shear and elongation. J. Rheol. 30:459–501 (1986).
37. F. J. Stadler, C. Piel, J. Kaschta, S. Rulhoff, W. Kaminsky, and H. Munstedt. Dependence of the zero shear-rate viscosity and the viscosity function of linear high-density polyethylenes on the mass-average molar mass and polydispersity. Rheol. Acta 45:755–764 (2006).
38. S. Saeda, J. Yotsuyanagi, and K. Yamaguchi. The relation between melt flow properties and molecular weight of polyethylene. J. Appl. Polym. Sci. 15:277–292 (1971).
39. A. Rudin and K. K. Chee. Zero shear viscosities of narrow and broad distribution polystyrene melts. Macromolecules 6:613–624 (1973).
40. D. I. W. Posch. Polyolefins. In: M. Kutz, ed. Applied Plastics Engineering Handbook – Processing, Materials, and Applications, 3rd ed., Elsevier, 2024, 29–56.
41. R. A. Bubeck. Structure-property relationships in metallocene polyethylenes. Mater. Sci. Eng. R 39:1–28 (2002).
42. A. Malmberg, C. Gabriel, T. Steffl, H. Munstedt, and B. Lofgren. Long-chain branching in metallocene-catalyzed polyethylenes investigated by low oscillatory shear and uniaxial extensional rheometry. Macromolecules 35:1038–1048 (2002).
43. W. W. Graessley. Effect of long branches on the flow properties of polymers. Accounts Chem. Res. 10:332–339 (1977).
44. C. J. Hawker, P. J. Farrington, M. E. Mackay, K. L. Wooley, and M. J. Frechet. Molecular ball bearings? The melt viscosity of dendritic macromolecules. J. Am. Chem. Soc. 117:4409–4410 (1995).
45. R. K. Gupta. Polymer and Composite Rheology. 2nd ed., Marcel Dekker, New York, 2000.
46. H. Munstedt. Rheological properties and molecular structure of polymer melts. Soft Matter 7:2273–2283 (2011).
47. R. A. Mendelson. Flow properties of polyethylene melts. Polym. Eng. Sci. 9:350–355 (1969).
48. R. A. Mendelson, W. A. Bowles, and F. L. Finger. Effect of molecular structure on polyethylene melt rheology. I. Low-shear behavior. J. Polym. Sci.: Part A-2 8:105–126 (1970).
49. J. Miltz and A. Ram. Flow behavior of well-characterized polyethylene melts. Polym. Eng. Sci. 13:273–279 (1973).
50. G. L. Wilkes. An overview of the basic rheological behavior of polymer fluids with an emphasis on polymer melts. J. Chem. Education 58:880–892 (1981).

51. G. Kraus and J. T. Gruver. Rheological properties of multichain polybutadienes. J. Polym. Sci.: Part A 3:105–122 (1965).
52. V. R. Raju, H. Rachapudy, and W. W. Graessley. Properties of amorphous and crystallizable hydrocarbon polymers. IV. Melt rheology of linear and star-branched hydrogenated poly butadiene. J. Polym. Sci.: Polym. Phys. Ed. 17: 1223–1235 (1979).
53. J. M. Carella. Comments on the paper "Comparison of the rheological properties of metallocene-catalyzed and conventional high-density polyethylenes." Macromolecules 29:8280–8281 (1996).
54. D. R. Saini and A. V. Shenoy. Viscoelastic properties of linear low density polyethylene melts. Eur. Polym. J. 19:811–816 (1983).
55. T. Masuda, Y. Ohta, and S. Onogi. Rheological properties of anionic polystyrenes. III. Characterization and rheological properties of four-branch polystyrenes. Macromolecules 4:763–768 (1971).
56. M. L. Williams, R. F. Landel, and J. D. Ferry. The temperature dependence of relaxation mechanisms in amorphous polymers and other glass-forming liquids. J. Am. Chem. Soc. 77: 3701–3706 (1955).
57. J. D. Ferry. Viscoelastic Properties of Polymers. 3rd ed., Wiley, New York, 1980.
58. H. Schott. Dependence of activation energy for viscous flow of polyhydrocarbons on bulk of substituents. J. Appl. Polym. Sci. 6:S29–S30 (1962).
59. R. S. Porter and J. F. Johnson. Temperature dependence of polymer viscosity. The influence of polymer composition. J. Polym. Sci.: Part C 15:373–380 (1966).
60. D. W. van Krevelen and P. J. Hoftyzer. Newtonian shear viscosity of polymeric melts. Angew. Makromol. Chem. 52:101–109 (1976).
61. M. J. Crochet and P. M. Naghdi. On thermomechanics of polymers in the transition and rubber regions. J. Rheol. 22:73–89 (1978).
62. D. A. Carey, C. J. Wust Jr., and D. C. Bogue. Studies in nonisothermal rheology: Behavior near the glass transition temperature and in the oriented glassy state. J. Appl. Polym. Sci. 25:575–588 (1980).
63. R. K. Gupta and A. B. Metzner. Modeling of nonisothermal polymer processes. J. Rheol. 26:181–198 (1982).
64. F. N. Cogswell. Polymer Melt Rheology. George Godwin, London, 1981.
65. M. H. Cohen and D. Turnbull. Molecular transport in liquids and glasses. J. Chem. Phys. 31:1164–1169 (1959).
66. P. B. Macedo and T. A. Litovitz. On the relative roles of free volume and activation energy in the viscosity of liquids. J. Chem. Phys. 42:245–256 (1965).
67. I. C. Sanchez. Towards a theory of viscosity for glass-forming liquids. J. Appl. Phys. 45:4204–4215 (1974).
68. F. N. Cogswell. The influence of pressure on the viscosity of polymer melts. Plastics and Polymers 41(Feb.):30–43 (1973).
69. R. F. Westover. Effect of hydrostatic pressure on polyethylene melt rheology. SPE Trans. 1:14–20 (1961).
70. J. C. McGowan and F. N. Cogswell. The effect of pressure and temperature upon the viscosity of liquids with special reference to polymeric liquids. Br. Polym. J. 4: 183–198 (1972).
71. M. R. Kamal and H. Nyun. The effect of pressure on the shear viscosity of polymer melts. Trans. Soc. Rheol. 17:271–285 (1973).
72. L. A. Utracki. A method of computation of the pressure effect on melt viscosity. Polym. Eng. Sci. 25:655–668 (1985).
73. S. Chakravorty, M. Rides, C. R. G. Allen, and C. S. Brown. Polymer melt viscosity increases under pressure: Simple new measurement method. Plast., Rubber Comp. Proc. Applic. 25:260–261 (1996).

74. H. E. Park and J. M. Dealy. Effects of pressure and supercritical fluids on the viscosity of polyethylene. Macromolecules 39:5438–5452 (2006).
75. R. A. Stratton. The dependence of non-Newtonian viscosity on molecular weight for "monodisperse" polystyrene. J. Colloid Interf. Sci. 22:517–530 (1966).
76. G. Kraus and J. T. Gruver. Melt viscosity of random copolymers of butadiene and styrene prepared by anionic polymerization. Trans. Soc. Rheol. 13:315–322 (1969).
77. R. L. Ballman, and R. H. M. Simon. The influence of molecular weight distribution on some properties of polystyrene melt. J. Polym. Sci.: Part A 2:3557–3575 (1964).
78. W. W. Graessley. Molecular entanglement theory of flow behavior in amorphous polymers. J. Chem. Phys. 43:2696–2703 (1965).
79. W. W. Graessley. Viscosity of entangling polydisperse polymers. J. Chem. Phys. 47:1942–1953 (1967).
80. E. van Ruymbeke, C.-Y. Liu, and C. Bailly. Quantitative tube model predictions for the linear viscoelasticity of linear polymers. Rheol. Rev. 53–134 (2007).
81. G. Marrucci and G. Ianniruberto. Flow-induced orientation and stretching of entangled polymers. Phil. Trans. R. Soc. Lond. A 361:677–688 (2003).
82. C. K. Shih. The effect of molecular weight and molecular weight distribution on the non-Newtonian behavior of ethylene-propylene-diene polymers. Trans. Soc. Rheol. 14:83–114 (1970).
83. V. C. Long, G. C. Berry, and L. M. Hobbs. Solution and bulk properties of branched polyvinyl acetates IV—Melt viscosity. Polymer 5:517–524 (1964).
84. R. A. Mendelson. Prediction of melt viscosity flow curves at various temperatures for some olefin polymers and copolymers. Polym. Eng. Sci. 8:235–240 (1968).
85. C. S. Brazel and S. L. Rosen. Fundamental Principles of Polymeric Materials. 3rd ed., Wiley, Hoboken, NJ, 2012.
86. A. Kumar and R. K. Gupta. Fundamentals of Polymer Engineering. 3rd ed., CRC Press, Boca Raton, FL, 2019.
87. E. H. Immergut and H. F. Mark. Principles of plasticization. Adv. Chem. 48:1–26 (1965).
88. L. E. Nielsen, and R. F. Landel. Mechanical Properties of Polymers and Composites. 2nd ed. Marcel Dekker, New York. 1994, p. 21.
89. A. D. Godwin. Plasticizers. In: M. Kutz, ed. Applied Plastics Engineering Handbook, 3rd ed., Elsevier, 2024, 595–618.
90. P. H. Daniels. A brief overview of theories of PVC plasticization and methods used to evaluate PVC-plasticizer interaction. J. Vinyl Addit. Technol. 15:219–223 (2009).
91. A. Bodaghi. An overview on the recent developments in reactive plasticizers in polymers. Polym. Adv. Technol. 31:355–367 (2020).
92. L. Slade and H. Levine. Water and glass transition – Dependence of the glass transition on composition and chemical structure: Special implications for flour functionality in cookie baking. J. Food Eng. 24:431–509 (1995).
93. M. Padmanabhan and M. Bhattacharya. Effect of extrusion processing history on the rheology of corn meal. J. Food Eng. 18:335–349 (1993).
94. K. F. Zieminski and J. E. Spruiell. Computer simulation of melt spinning. J. Appl. Polym. Sci. 35:2223–2245 (1988).

4 Shear Viscosity of Polymer Solutions

I. INTRODUCTION

The shear flow of polymer solutions, both dilute and concentrated, and employing either water or an organic liquid as the solvent, is encountered in a wide variety of technologically important situations. Common examples include the addition of 2–3% of a polymer such as polymethyl methacrylate to a mineral oil in order to formulate multigrade motor oils; the polymer serves to reduce the dependence of oil viscosity on temperature and also to lower the pour point. More concentrated solutions, containing up to 30 or 40% polymer, are used to manufacture textile fibers by the process of (wet or dry) solution spinning. This is because, though polymers such as cellulose acetate cannot be melted, they can be dissolved in a solvent such as acetic acid or acetone. In the petroleum industry, aqueous solutions of xanthan gum polysaccharides and partially hydrolyzed polyacrylamides in a concentration range of a few hundreds of ppm are utilized for enhanced oil recovery by water flooding; the thickening effect of the added polymer reduces the mobility difference between the injected water and the oil in place, thereby improving both areal and vertical sweep efficiencies. The thickening effect of polymers, generally referred to as *hydrocolloids* or gums and which can include *polysaccharides* and *proteins*, is also put to good use in the food industry [1]. Common sources of polysaccharides are botanical (plants, seeds, and trees), algal (various seaweeds), and microbial, while proteins may be derived from plants and animals. For example, xanthan gum is a common additive in salad dressings and other products. In food emulsions, polysaccharides can reduce interfacial tension between the two phases, and the result is a fine dispersion that is also stable. Paints and coatings are other applications of polymer solutions. Note that high-molecular-weight polymers can be fairly shear-sensitive, and chain scission takes place quite easily. It is for this reason that, to achieve a desired enhancement in viscosity, one often adds a large amount of low-molecular-weight polymer instead of a small amount of high-molecular-weight polymer to a solvent if large deformation rates are likely to be encountered.

We obviously expect the viscosity of a polymer solution to depend on concentration. In addition, as in the case of a polymer melt, we expect that the viscosity of a polymer dissolved in water or in an organic solvent will be a function of temperature, pressure, shear rate, molecular weight, molecular weight distribution, and molecular architecture (including chain stiffness). Typical viscosity–shear rate curves for polystyrene solutions of different concentrations are shown in Figure 4.1 [2]; polystyrene is a flexible chain polymer, and the concentrations are such that polymer–polymer entanglements are present in each instance. These curves show that dilute solutions remain Newtonian in behavior to higher rates of shear than do more concentrated solutions. This is because the presence of the solvent reduces the number of chain

DOI: 10.1201/9781003013501-4

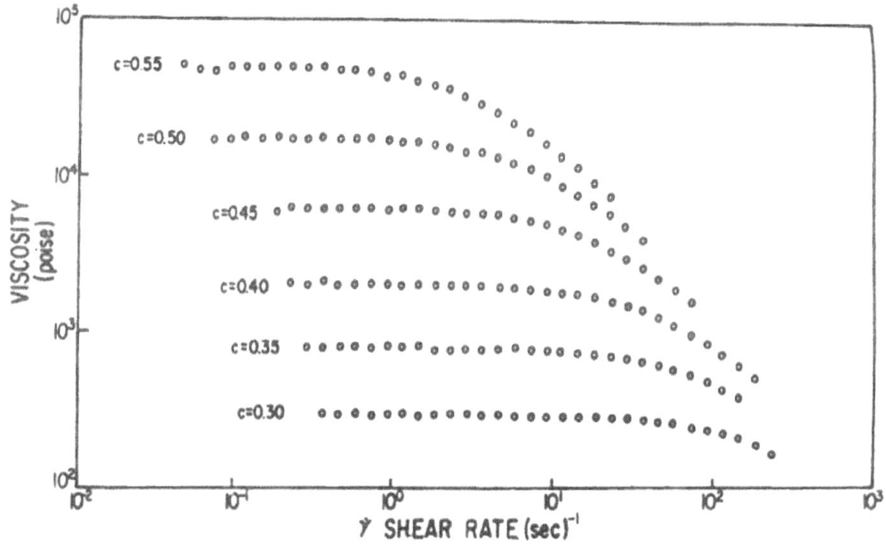

FIGURE 4.1 Shear rate dependence of polystyrene solutions in *n*-butyl benzene at different concentrations (g/cc) at 30°C. Molecular weight = 411,000. (From Ref. [2].)

entanglements per unit volume. The absolute lower limit of the solution viscosity is of course the solvent viscosity; for the solvent used in Figure 4.1, this is 1.3 centipoise. Since the various curves in Figure 4.1 are qualitatively similar to the corresponding curves in Figure 3.1, they can again be represented using the power-law model or the Carreau model, Eqs. (3.2) and (3.3), respectively. If a more viscous solvent is used, the solution viscosity does tend to approach the solvent viscosity at high enough shear rates, giving an upper Newtonian region. Now the Carreau equation is written as

$$\frac{\eta_a - \eta_\infty}{\eta_0 - \eta_\infty} = \left[1 + \left(\lambda_c \dot{\gamma}\right)^2\right]^{(n-1)/2} \tag{4.1}$$

in which η_∞ is the limiting viscosity at high shear rates, usually taken to be the solvent viscosity. Xanthan gum is considered to have a more rigid backbone as compared to polystyrene, but it exhibits qualitatively similar flow behavior in aqueous, salt-free solutions; flow curves, as a function of polymer concentration, can be described using the Cross model, Eq. (3.5), that has been appropriately modified to account for the solvent viscosity [3]. The shear-thinning region can again be described by the power-law model. If the solvent is extremely viscous, the solution and solvent viscosities may be almost the same, independent of the shear rate. This situation results in a constant viscosity, elastic liquid [4, 5].

The treatment in this chapter parallels that of the previous chapter in that the influence of each process and structural variable on the viscosity of polymer solutions is examined in a systematic manner. While polymer concentration is a new variable, additional variables arise if the polymer contains ionizable groups. In a high dielectric medium such as water, a polymer such as partially hydrolyzed polyacrylamide can ionize to give a polyion that assumes an extended conformation due to electrostatic repulsion between charged groups. The presence of salt, however, tends to screen these charges resulting in a more coiled conformation. Changes in conformation lead to changes in viscosity. Water-soluble polymers can also be modified by the incorporation of hydrophobic moieties. These can associate with each other, resulting in the formation of a network that again alters the solution viscosity. Note that water-soluble polymers are also commonly employed as flocculants in water treatment, and block copolymers are used to stabilize a solid-in-liquid suspension sterically; these applications are considered in the chapter on the rheology of suspensions.

II. INTRINSIC VISCOSITY—THE LIMIT OF INFINITE DILUTION

If we think of an infinitely dilute polymer solution to be a suspension of isolated, noninteracting spheres in a Newtonian liquid, we can immediately relate the *relative viscosity* η_R to the polymer volume fraction ϕ through the *Einstein result* [6]:

$$\eta_R = \frac{\eta}{\eta_s} = 1 + 2.5\phi \tag{4.2}$$

in which η and η_s are the solution and solvent viscosities, respectively. The polymer volume fraction, in turn, can be related to the mass concentration c (see Ref. [7], for example) to yield

$$\phi = \frac{vcN_A}{M} \tag{4.3}$$

where v is the volume occupied by a single polymer molecule, N_A is Avogadro's number, and M is the polymer molecular weight. In writing Eq. (4.3), it is assumed that the polymer sample is monodisperse and that each molecule assumes a spherical shape in solution. Introducing Eq. (4.3) into Eq. (4.2), rearranging the result, and extrapolating to zero concentration gives the *intrinsic viscosity* or the *limiting viscosity number* $[\eta]$. Extrapolation is needed because measurements are made on dilute solutions, and there may be hydrodynamic interactions between neighboring molecules. In other words, the flow around one molecule may be influenced by the presence of other molecules. Thus,

$$[\eta] = \lim_{c\to 0} \frac{\eta_R - 1}{c} = 2.5\frac{vN_A}{M} \tag{4.4}$$

Experimental data cannot be expected to obey Eq. (4.4) because polymer molecules are not rigid spheres. Rather, they tend to entrap solvent molecules, and one finds that data can generally be represented in terms of the *Huggins* equation:

$$\frac{\eta_R - 1}{c} = [\eta] + k[\eta]^2 c \tag{4.5}$$

where k is known as the Huggins constant. Eq. (4.4) shows that the intrinsic viscosity is directly related to the size of a polymer molecule in solution. If the coil size changes due, say, to changes in temperature, pH, or ionic strength, the intrinsic viscosity must also change. Furthermore, if the coil size or sphere volume can be related to the polymer molecular weight, one can employ Eq. (4.4) for the measurement of the molecular weight. This connection between the sphere radius r and the polymer chain length is usually made with the help of a molecular model. If, for example, one takes a linear, flexible polymer to be freely jointed chain of n links, each of length l, one can show that [7]

$$r \propto \sqrt{n}\,l \tag{4.6}$$

provided there are no excluded volume effects that are long-range interactions due to attraction and repulsion forces between widely separated chain segments or between polymer segments and solvent molecules. This happens at what is known as the *theta condition* [8].

Because the polymer molecular weight is proportional to n and the sphere volume, v, is proportional to the third power of r, a combination of Eqs. (4.4) and (4.6) reveals that

$$[\eta] \propto M^{1/2} \tag{4.7}$$

which is valid under theta conditions alone; these generally arise when a poor solvent is used. In a good solvent, polymer–solvent interactions result in coil expansion [9], and we find that

$$[\eta] = KM^{a} \tag{4.8}$$

which is known as the *Mark–Houwink–Sakurada* equation. The values of the constants K and a depend on the polymer, the solvent, and the temperature; these are determined experimentally using monodisperse polymer fractions, and extensive tabulations are available [9, 10]. The value of the exponent a typically varies from 0.5 at the theta temperature to 0.8 for a very good solvent. However, the value of a increases beyond 0.8 as the chain stiffness increases, and for rigid-rod polymers, a reaches a value of 1.8; this result can be obtained theoretically by assuming that polymer molecules take the form of long cylinders in solution [11]. It can also be understood intuitively if we say that for a rigid polymer chain, r is proportional to n. Thus, the hydrodynamic volume of the polymer is now proportional to M^3. Consequently, Eq. (4.4) suggests that $[\eta]$ is proportional to the square of the molecular weight [12]. Note that we have implicitly assumed that the intrinsic viscosity does not depend on the shear rate. This happens when the tendency of a polymer molecule to extend and orient with the flow field is counteracted by Brownian motion. Thus, the quantity defined by Eq. (4.8) is actually the zero-shear-rate intrinsic viscosity, $[\eta]_0$. It is found that $[\eta]$ typically exhibits a Newtonian plateau at low shear rates, and this is followed by a shear-thinning region. According to Eqs. (4.5) and (4.8), $[\eta]_0$ should increase with increasing molecular weight at a fixed mass concentration.

The obvious utility of intrinsic viscosity measurements is that they afford a simple way of polymer molecular weight measurement. In addition, the measurement of intrinsic viscosity can reveal the occurrence of order–disorder transitions. This is best illustrated by the behavior of xanthan gum, an anionic polysaccharide produced by the bacterium *Xanthomonas* campestris and having a trisaccharide side chain on every second residue; the polymer molecular weight ranges from 1 to 50 × 10^6 g/mol. Xanthan gum dissolves easily in water, and, at room temperature, it takes an extended chain conformation due to electrostatic repulsion between the charges on the side chains. If the temperature is raised, the molecules assume a coil-like shape. If salt is added to the solution at room temperature, a helical conformation results because the side chains collapse onto the backbone due to charge screening. Upon heating, the molecules again take on the shape of coils, but the transition takes place at a temperature that is higher than in the absence of salt [13]. One, therefore, finds that flow curves show distinct differences upon changing salt concentration, cation type, and temperature. These changes have been investigated by carrying out intrinsic viscosity measurements and have been related to the shape and size of the polymer molecule in solution [14].

In terms of actually measuring the relative viscosity as a function of concentration for the purpose of computing the intrinsic viscosity, any of the viscometers described in Chapter 2 can be used, provided that the shear rate is low enough that we are in the lower Newtonian region. This requirement is generally satisfied if the solutions are so dilute that the relative viscosity does not appreciably exceed unity. In such a case, it is customary to employ (inexpensive) glass capillary viscometers of the type shown in Figure 4.2. This particular instrument is known as a suspended level *Ubbelohde* viscometer. In use, one fills bulb A with the polymer solution, allows the liquid to drain through the capillary under the influence of gravity, and notes the

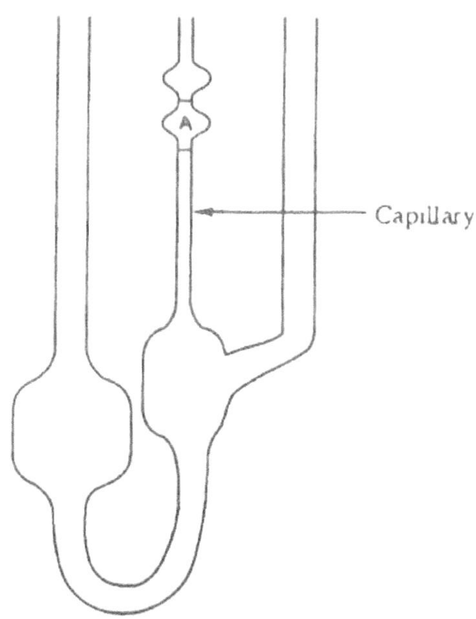

FIGURE 4.2 An Ubbelohde viscometer. (From Ref. [15].)

time taken for the liquid level to move between the two marks made on bulb A. The process is repeated for the pure solvent and also for the polymer solution at different concentrations. Note that proper cleaning of the capillary and good temperature control are required to obtain consistent data.

If we recall that the flow of a Newtonian liquid through a capillary is described by the Hagen–Poiseuille equation [Eq. (2.5)], then it is evident that the relative viscosity is given by the ratio of the flow rate of the pure solvent to that of the solution. The volume of bulb A, however, is fixed, and the flow rate is inversely proportional to the efflux or drainage time t, with the result that

$$\frac{\eta}{\eta_s} = \frac{t\rho}{t_s\rho_s} \tag{4.9}$$

and we may calculate the relative viscosity with ease, knowing the densities of the solution, ρ, and the solvent, ρ_s. If it is found that the shear rate is not low enough, extrapolation of the data to lower shear rates may be necessary. Also, if the polymer sample is not monodisperse, the molecular weight calculated with the help of Eq. (4.8) will be an average value intermediate between the number-average molecular weight and the weight-average molecular weight. Such data can still be useful for quality control purposes.

In the above treatment, polymer molecules were assumed to be spheres. However, many natural and biological polymers are non-spherical, and the effect of particle shape as well as particle porosity has been studied [16]. Other variables that have been examined are chain branching and solvent quality. Finally, results are available for the intrinsic viscosity of polyelectrolyte solutions and liquid-crystal polymers in nematic solvents [16].

In closing this section, we note that Saini and Shenoy [17] have proposed a simple method for the generation of temperature-independent master curves for polymer solution viscosity as a function of shear rate using data obtained with the help of glass capillary viscometers of the type considered here.

III. ZERO-SHEAR VISCOSITY AS A FUNCTION OF MOLECULAR WEIGHT AND CONCENTRATION

The zero-shear viscosity of a polymer–solvent system can change quite considerably in going from pure solvent to undiluted polymer. In the extreme case of a low-viscosity liquid added to a polymer near its glass transition temperature, the viscosity may change by a factor of 10^{15} in covering this concentration range. Even at temperatures well above T_g, the viscosity of a polymer solution has been found to vary from the first power to greater than the 14th power of the concentration of the polymer, depending upon the concentration range, the polymer–solvent system, temperature, and other experimental conditions [18]. Typical experimental data are shown in Figure 4.3 [19].

In terms of obtaining a rough estimate of the viscosity of a polymer solution at a temperature well above T_g, the data of Figure 4.3 suggest the following mixture rule:

$$\log\eta_0 = \phi_p \log\eta_p + \phi_L \log\eta_L \tag{4.10}$$

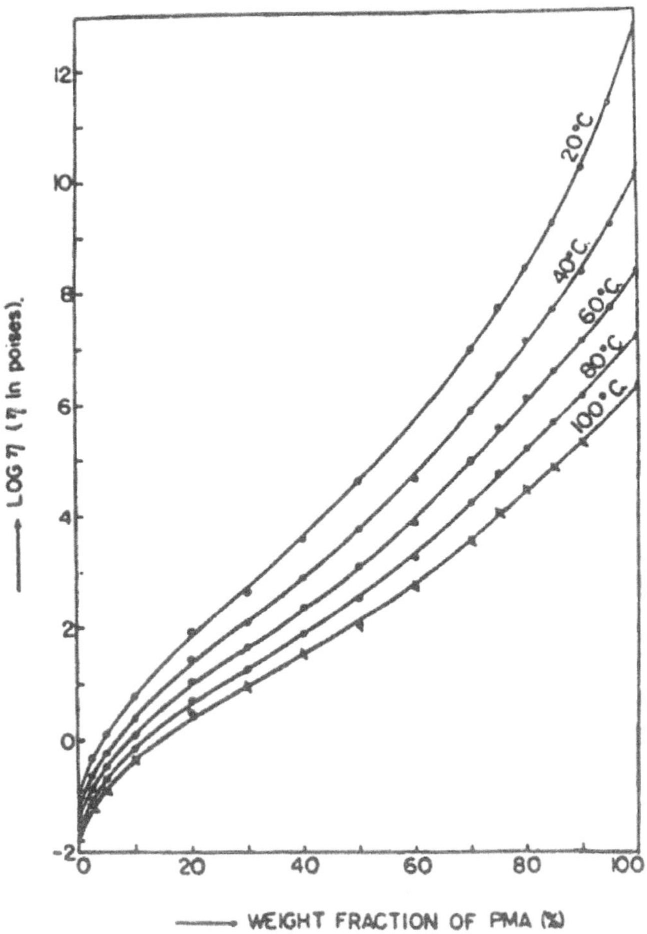

FIGURE 4.3 Viscosity of polymethyl acrylate dissolved in diethyl phthalate. (Reprinted with permission from Ref. [19]. Copyright 1962 American Chemical Society.)

The viscosities of the pure polymer and the liquid are η_P and η_L, respectively; ϕ_P and ϕ_L are the corresponding volume fractions. The difficulty in finding a single (empirical) equation that can accurately represent data over the entire concentration and temperature range is related to the fact that the physics of the situation changes as the polymer concentration changes. In the limit of infinite dilution, long-chain polymer molecules behave in the fashion of impenetrable spheres, while as bulk polymer we have an entangled mass of fluid. At an intermediate concentration, the spheres bump up against each other, and there is coil overlap but no entanglement. These three regions are roughly differentiated by the value of the product $[\eta]c$ [16]. According to Eq. (4.4), $[\eta]M/N_A$ is the volume of a polymer molecule multiplied by 2.5, whereas cN_A/M is obviously the number of polymer molecules per unit volume. As a consequence, $[\eta]c$, which is the product of these two quantities, represents the volume fraction of polymer multiplied by 2.5. If this number is small compared to unity, the polymer solution is considered dilute; if in the range of 1–10, the solution is semidilute or low to moderately concentrated; if above 10, the solution is entangled.

Based on the foregoing picture, it is reasonable to expect that different theories and correspondingly different equations will portray polymer solution viscosity data in the different concentration regimes. We have already seen that the Huggins equation [Eq. (4.5)] fits data in the dilute regime. It is found that data in the coil-overlap region can often be represented with the help of the *Martin* equation [20]:

$$\frac{\eta_{sp}}{[\eta]c} = \exp\left(k[\eta]c\right) \tag{4.11}$$

in which η_{sp} is the *specific viscosity*, defined as $(\eta_R - 1)$. A feature of both the Huggins equation and the Martin equation is that the relative viscosity is postulated to be a unique function of $[\eta]c$. This observation was exploited by Lyons and Tobolsky [21], who proposed an empirical equation for the viscosity of polymer solutions over the entire concentration range, provided that the molecular weight of the polymer was below the value needed for entanglement formation. In the Lyons–Tobolsky equation, since the relative viscosity is defined with respect to the solvent viscosity, the accuracy of the theory is expected to be poor for extremely concentrated solutions [22]. However, the equation has been found to be satisfactory in a number of cases [23]. The Lyons–Tobolsky equation is as follows:

$$\frac{\eta_{sp}}{[\eta]c} = \exp\frac{\left(k[\eta]c\right)}{1 - bc} \tag{4.12}$$

A new constant, b, appears in Eq. (4.12), and it can be considered an empirical constant, or it can be calculated from melt data on the pure polymer.

Besides empirical equations, there are several fundamental theories that seek to relate the zero-shear viscosity of a polymer solution to the mass concentration and the polymer molecular weight. If the polymer concentration is such that there are no chain entanglements present, an explicit expression can be derived for the zero-shear viscosity by examining the behavior of a single, isolated polymer molecule.

In the *Bueche theory* [24], the energy dissipated per unit volume due to fluid friction in a shear field is calculated by multiplying the energy dissipated by the motion of a single polymer molecule times the number of polymer molecules per unit volume. This quantity then leads to the following expression for the polymer contribution to the zero-shear viscosity:

$$\eta_0 = \frac{cN_A\zeta nl^2}{36M_0} \tag{4.13}$$

in which c is the mass concentration, N_A is Avogadro's number, M_0 is the molecular weight of a monomer unit, ζ is the friction coefficient per monomer unit (similar in concept to Stokes' law for drag on spheres), and n and l are the same quantities as in Eq. (4.6). Since n is proportional to molecular weight M, the zero-shear viscosity is predicted to be proportional to the product cM.

In the bead–spring theories, each polymer molecule in solution is idealized as $(N + 1)$ spheres, each of mass m, connected by N massless springs [25]. The polymer solution then is a noninteracting suspension of these stringy entities in a Newtonian liquid. Under the influence of flow, the springs get stretched and oriented, and the tension in the spring makes a contribution to the stress in the liquid. By adding together the contributions from all the (identical) polymer molecules, we get the complete stress, from which the viscosity can be determined, and, not surprisingly, one finds a linear relationship between η_0 and (cM). In the *Rouse model*, the polymer molecules are taken to be free draining (not impenetrable), and the viscosity contributed by the polymer molecules under theta conditions is found to be [26]

$$\eta_0 = \frac{cMN_A l^2 \zeta}{36 M_0^2} \tag{4.14}$$

Again, the zero-shear viscosity is predicted to increase linearly with cM.

When experimental data for the zero-shear viscosity of polymer solutions are examined, it is found that the dependence of the viscosity on cM is stronger than the linear relation predicted by Eqs. (4.13) and (4.14). It turns out that this is due to the variation of the friction coefficient ζ with polymer concentration. Note that changes in polymer concentration lead to changes in the free volume. When data are corrected to give viscosities at a constant value of the monomeric friction coefficient (or equivalently the same free volume), linear behavior is established. This is shown in Figure 4.4 [27]. An examination of Figure 4.4 also reveals that linearity between the viscosity and cM persists only up to a critical value of cM. Beyond this critical value, the slope of the viscosity-versus-cM curve increases to about 3.4 on logarithmic coordinates. The behavior of polymer solutions, therefore, is entirely analogous to that of polymer melts, for which c equals the melt density. The change in slope in Figure 4.4 is again due to the formation of chain entanglements. The critical molecular weight for entanglement formation in solution now depends on the polymer concentration and is given by

$$(Mc)_{\text{sol}} = \left[\frac{\rho}{c}\right] Mc \tag{4.15}$$

where Mc is the molecular weight needed for entanglement formation in the melt. Behavior similar to that shown in Figure 4.4 is observed for polysaccharide solutions when η_{sp} is plotted versus $c[\eta]$ and can be represented by a master curve for several, but not all, polysaccharides [28].

In order to theoretically explain the viscosity behavior beyond this critical value of cM, it is fruitful to consider the dilute region to be followed by a semi-dilute region and a concentrated region. It is found that the result of applying the scaling

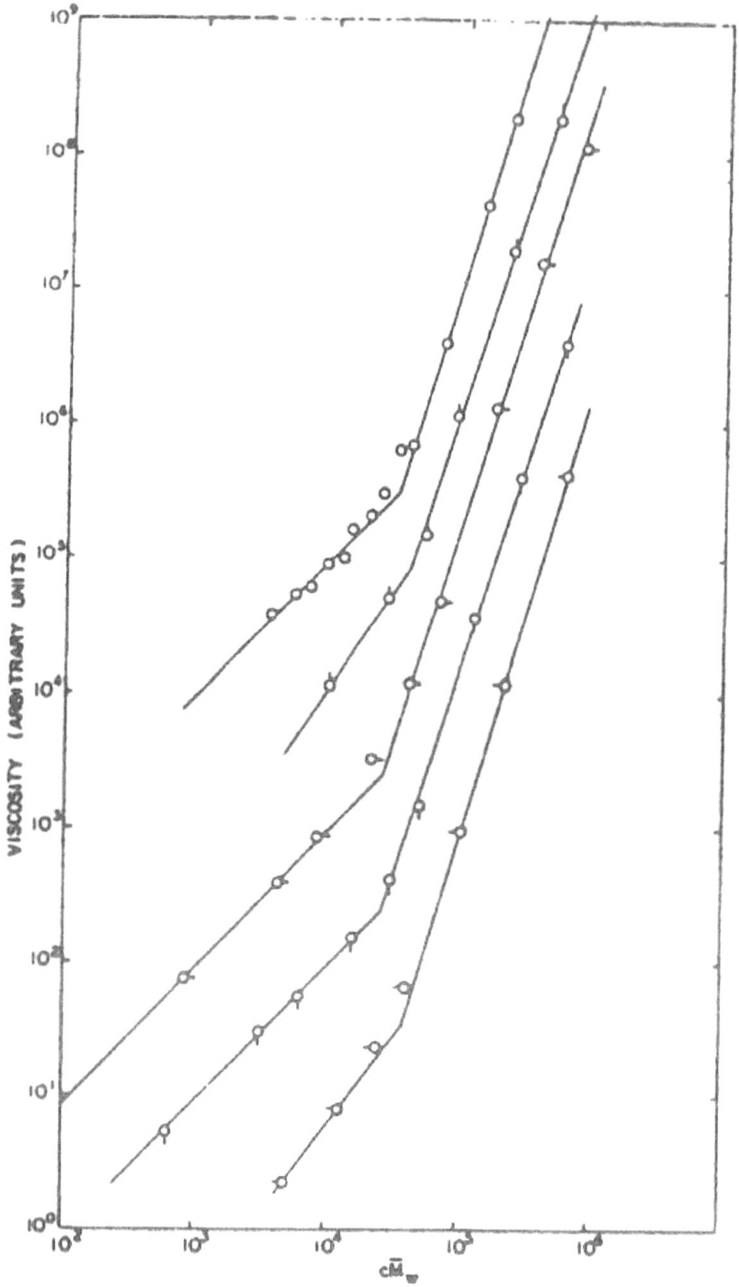

FIGURE 4.4 Viscosity versus the product *cMw* for polystyrene. The uppermost curve is for the molten polymer at 217°C. Progressively lower curves are for 0.55 g/ml polymer in *n*-butyl benzene, 0.415 g/ml and 0.31 g/ml in di-octyl phthalate, and 0.255 g/ml in *n*-butyl benzene. Data at the various concentrations have been shifted vertically to avoid overlap. (From Ref. [27].)

theory of de Gennes to the prediction of the zero-shear reduced viscosity leads to good agreement with experimental results without the use of any adjustable parameters. In the concentrated regime, however, one needs to assume that the zero-shear viscosity is proportional to $(cM)^{3.4}$ multiplied by a friction coefficient whose value depends on polymer concentration (see [29] and [16] and references therein).

The idea that the viscosity of a polymer solution is determined by the free volume has been employed by Kelley and Bueche [30] to predict the viscosity of concentrated (entangled) polymer solutions using the viscosity of the pure polymer as the reference. The final equations are as follows:

$$\frac{\eta}{\eta_p} = \phi_p^4 \exp\left(\frac{1}{\phi_p f_p + \phi_L f_L} - \frac{1}{f_p}\right) \tag{4.16}$$

$$f_p = 0.025 + 4.8 \times 10^{-4}\left(T - T_{gp}\right) \tag{4.17}$$

$$f_L = 0.025 + \alpha_L\left(T - T_{gL}\right) \tag{4.18}$$

The free volumes of the polymer and the liquid are f_p and f_L, respectively. The difference between the coefficients of expansion of the liquid above and below T_g is α_L. Generally, α_L is of the order of 10^{-3}. The glass transition temperatures of the polymer and solvent are T_{gp} and T_{gL}, respectively. Figure 4.5 shows a plot of Eq. (4.16) for several values of f_p and f_L. The figure shows the tremendous sensitivity of the relative viscosity to small changes in f_p and f_L. The concentration dependence of relative viscosity is especially great at temperatures close to the T_g of the pure polymer, as represented by curve D. Very small amounts of a liquid or plasticizer can reduce the viscosity of the polymer by one or more decades. Eqs. (4.17) and (4.18) give approximate theoretical values of f_p and f_L. However, because of the extreme sensitivity of viscosity to the free volumes, in practical situations it may be more convenient to consider f_p and f_L as empirical constants. Also, since Eq. (4.16) was derived for entangled macromolecules, it may not hold for dilute polymer solutions.

Many natural polymers, including polysaccharides, are stiff-chain, rodlike polymers, and their behavior in dilute solution is similar to that of flexible chain polymers, i.e., the shear viscosity is a linear function of cM. This behavior can be predicted theoretically if one considers the polymer solution to be made up of rigid dumbbells suspended in the solvent [25]. With increasing concentration, though, a situation is reached where the rotational motion of one molecule is hindered by the presence of another molecule. This happens when the number of polymer molecules per unit volume equals $1/L^3$, where L is the length of the polymer molecule. Ultimately, at a still higher concentration of $1/(dL^2)$, where d is the rod diameter, random orientation of molecules is not permitted, and an ordered, liquid crystalline phase is formed. Between these two concentrations, it is found that the zero-shear viscosity increases with the fifth to eighth power of concentration and the seventh to eighth power of molecular weight [11].

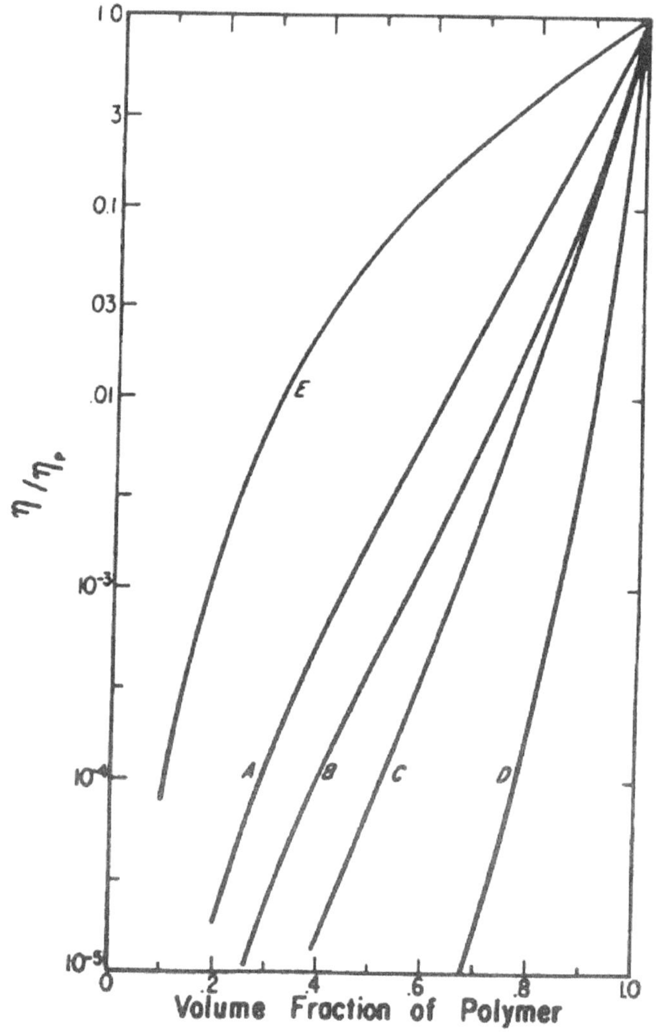

FIGURE 4.5 Viscosity of solutions divided by the viscosity of the polymer at the same temperature as a function of the concentration of the polymer according to the theory of Kelley and Bueche. Values of f_P and f_L, respectively, for the different curves are A: 0.1, 0.2; B: 0.1, 0.3; C: 0.05, 0.1; D: 0.05, 0.2; E: $f_p = f_L$. (From Ref. [32].)

IV. INFLUENCE OF STRUCTURAL FACTORS, TEMPERATURE, AND PRESSURE ON THE ZERO-SHEAR VISCOSITY

The influence of polydispersity on the zero-shear viscosity of polymer solutions is the same as the influence of this variable on the zero-shear viscosity of polymer melts (seen earlier in Section II.A of Chapter 3). In other words, viscosity data on polydisperse polymers superpose with the data shown in Figure 4.4 when the weight-average molecular weight is used in place of M_w. The effect of chain branching is similar (see Section II.B of Chapter 3). In a given solvent, the branched polymer has a smaller radius of gyration and a smaller intrinsic viscosity than a linear polymer of the same molecular weight. At low and intermediate concentrations of polymer in the same solvent, there are no chain entanglements, and the viscosity of the branched polymer is lower than that of the linear polymer; it is found that [31]

$$\eta_B(c,M) = \eta_L(c,gM) \tag{4.19}$$

where g is the square of the ratio of the branched to unbranched radii of gyration, and the subscripts B and L denote branched and linear, respectively. At high concentrations, deviations from Eq. (4.19) begin to occur when the branches are long enough to form entanglements. Now, measured values of the viscosity of branched polymers can be as much as 100 times the viscosity of linear polymers at comparable c, gM values [31]. In terms of reptation, viscosity increases because the entangled branches prevent the entire chain from moving inside the confining tube. It is first necessary for the arms to retract and reinsert themselves into different tubes in new directions [16].

The temperature dependence of the zero-shear viscosity of polymer solutions follows the WLF equation [Eq. (3.9)], at low temperatures and the Arrhenius equation [Eq. (3.10)] at high temperatures. Recall that the WLF equation involves the polymer glass transition temperature as the reference temperature. For polymer solutions, this quantity can be estimated with the help of Eq. (3.17), or it may be measured using dilatometry, since the specific volume of polymers and their solution changes slope on going through the glass transition temperature [33]. Note that if a different temperature is used as the reference temperature, the appropriate form of the WLF equation is Eq. (3.19). Reference [34] may be consulted for additional details.

The pressure dependence of the viscosity of polymer solutions is important in a number of applications, including elastohydrodynamic lubrication, where the pressure often exceeds 1 GPa [35]. Bair and Winer [36] constructed a falling-body viscometer to measure the high-temperature, high-pressure viscosity of a number of liquid lubricants; they found that the isothermal viscosity could increase by several orders of magnitude upon increasing the pressure from atmospheric to 0.5 GPa, especially at lower temperatures. These data were explained by Yasutomi et al. [35], who modified the WLF equation to include the effect of pressure on the glass transition temperature and the free volume. The modified form of the WLF equation is

$$\log \frac{\eta_0(T,p)}{\eta_0(T_g)} = \frac{c_1\left[T - T_g(p)\right]F}{c_2 + \left[T - T_g(p)\right]F} \tag{4.20}$$

where the viscosity at the glass transition temperature is taken to be independent of pressure. The glass transition temperature itself depends on pressure according to

$$T_g = T_{g0} + A_1 \ln(1 + A_2 p) \tag{4.21}$$

while the free-volume expansion coefficient F is given by

$$F = 1 - B_1 \ln(1 + B_2 p) \tag{4.22}$$

in which A_1, A_2, B_1, and B_2 are constants. For a given liquid, these constants have to be determined by a regression analysis of sufficient viscosity–pressure data. Wu et al. [37] further developed the theory and demonstrated good agreement with data on blends of polymers with mineral oils.

V. SHEAR RATE DEPENDENCE OF VISCOSITY

The flow curve of a polymer solution depends on polymer concentration, polymer molecular weight, solvent viscosity, temperature, and the applied shear rate. The influence of concentration and shear rate on the measured viscosity is shown in Figure 4.1. This figure is qualitatively similar to Figures 3.1 and 3.10, which show the influence of temperature and molecular weight, respectively, on the curve of melt viscosity versus shear rate. Consequently, it is not surprising that, for a given polymer, data superposition that is independent of concentration, molecular weight, solvent viscosity, and temperature can be obtained by plotting the ratio of the solution viscosity to the zero-shear viscosity, essentially a relative viscosity, in terms of an appropriate dependent variable. In the simplest case, we can simply plot the relative viscosity against $\dot{\gamma} / \dot{\gamma}_0$, where $\dot{\gamma}_0$ represents the shear rate at which the viscosity has fallen to 90% of its zero-shear value. Tam and Tiu [38] showed that this procedure worked satisfactorily for aqueous solutions of both flexible chain polymers such as polyacrylamide and polyethylene oxide, and semirigid polymers, such as xanthan gum and carboxymethyl cellulose; the same procedure has also been found to work for polysaccharides [28] with the flow curves collapsing into a master curve for multiple polysaccharides. If the solvent viscosity, η_s, is significant in comparison to the solution viscosity, better data superposition may be obtained by plotting $(\eta - \eta_s)/(\eta_0 - \eta_s)$. For predictive purposes, one needs to know $\dot{\gamma}_0$; this may be taken to be the reciprocal of a characteristic time λ_E defined as follows [39]:

$$\lambda_E = \frac{\eta_0 - \eta_s}{c} \frac{M}{RT} \tag{4.23}$$

which appears to work well for polyethylene oxide dissolved in water or in a mixture of water and glycerol.

For polymers that exist as isolated coils in solution, shear thinning is the result of change in shape from a sphere to an ellipsoid; under the influence of shear, there is a reduction in the cross-sectional area that the polymer molecule presents to the flow field, and this reduces the drag force and hence the viscosity. For more concentrated solutions, the situation is similar to the case of polymer melts, and the decrease in solution viscosity with increasing shear rate is considered to be the result of shear-induced reduction in the number of entanglements. It is assumed that a certain amount of time, of the order of the fluid relaxation time, is needed for a polymer chain to both disengage itself from other chains and get entangled with them. It is further assumed that a polymer chain can entangle itself only with those chains whose centers lie within a sphere of specified radius. Thus, a reduction in entanglement density comes about in a sheared fluid because some polymer molecules, which would otherwise be entangled with the chain under consideration, pass through this sphere of influence with a residence time that is less than the fluid relaxation time [40]. The variation of

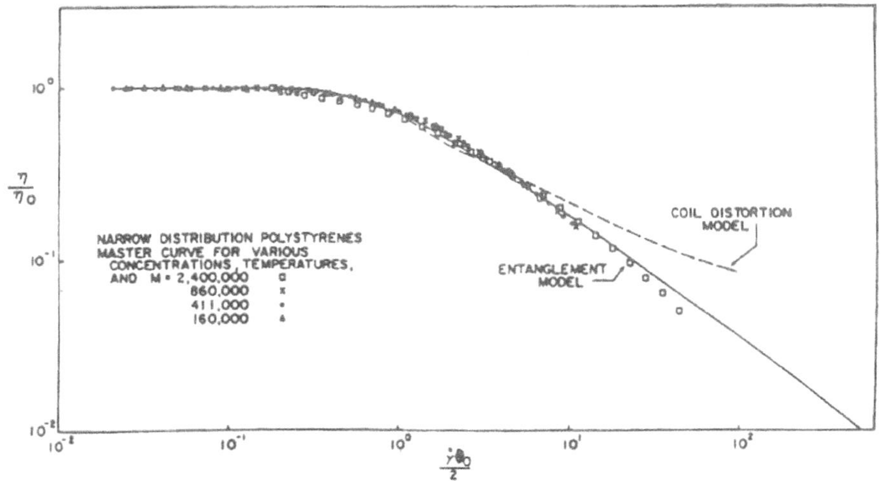

FIGURE 4.6 Master curve for polystyrene solutions in *n*-butyl benzene. The solid line represents Eq. (3.16). (From Ref. [2].)

the relative viscosity of a polymer solution with shear rate should, therefore, again be described by Eq. (3.16). This is indeed found to be true, and Figure 4.6 shows that a master curve is obtained when viscosity data on monodisperse polystyrenes for a variety of concentrations, molecular weights, and temperatures are plotted as suggested by this equation [2]; also included in Figure 4.6 are the data shown earlier in Figure 4.1. Since data on undiluted polymers too lie on the same master curve, the influence of polydispersity and chain branching on the non-Newtonian viscosity of polymer solutions is identical to that discussed for polymer melts in Section III.A of Chapter 3. While the idea of polymer chain entanglements has been replaced by the idea of reptation in the rheology literature, it remains a useful way of visualizing the behavior of polymer molecules subjected to shear.

VI. VISCOSITY OF POLYELECTROLYTES

Proteins and many other biological macromolecules are multiply charged compounds, as are some commonly encountered synthetic polymers, such as partially hydrolyzed polyacrylamide used in enhanced oil-recovery operations. Furthermore, by adding charged groups to a synthetic polymer such as polystyrene, we can obtain a water-soluble polymer, polystyrene sulfonate in this particular case. The rheological behavior of these *polyelectrolytes* in solution can be vastly different from that of the uncharged polymers considered thus far [41, 42]. This is because of interactions between different segments of a given polymer molecule, between different molecules, and between polymer and solvent. For example, when salt is added to xanthan gum in aqueous solution, the solution viscosity decreases because the molecules take on an ordered helical shape. Then, however, the viscosity becomes insensitive to further increases in salt concentration because the conformation of molecules is not affected even at very high salt contents [13]. Other interactions, especially at low polymer concentrations, alter the apparent size of the polymer in solution as determined by intrinsic viscosity measurements. A polyion can dissociate in a solvent such as water, and this results in like charges being distributed along the length of the polymer molecule. Electrostatic repulsion among the like charges leads to coil expansion and to an increase in the solution viscosity relative to the unexpanded coil. As the polymer concentration is increased, the solution viscosity tends to increase, because the number of coils increases. However, due to the presence of like charges, the coils repel each other, and their size decreases even as their number increases. Consequently, one finds that as the polymer concentration increases, the overall solution viscosity increases at a slower rate compared to uncharged polymers. Fuoss [43] found that

$$\eta - \eta_s \sim c^{1/2} \tag{4.24}$$

as opposed to the linear increase predicted by Eq. (4.4). This effect can be reversed by the addition of salt or another electrolyte that shields the charges. If a sufficient amount of salt is added, the charged macromolecule can even assume a random coil conformation and behave similarly to a nonionic polymer in a theta solvent. This behavior carries over to the entangled regime as well, where the zero-shear viscosity increases with the 3/2 power of the concentration instead of the 3.4 power. This is illustrated in Figure 4.7 by the data of Fernandez Prini and Lagos on aqueous

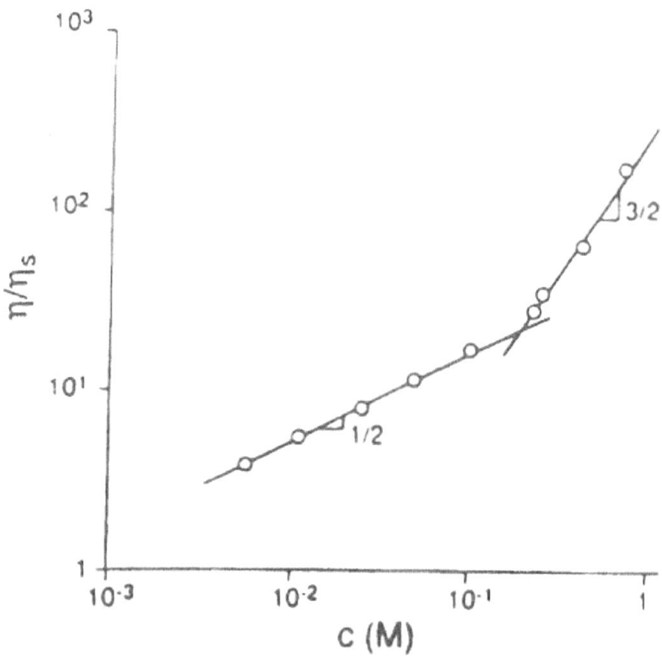

FIGURE 4.7 Concentration dependence of the relative viscosity for sodium polystyrene sulfonate in the absence of added salt. (Reprinted with permission from Ref. [46]. Copyright 1995 American Chemical Society.)

solutions of polystyrene sulfonate with sodium counterions [44]. These observations, and also the change in the zero-shear viscosity with polymer concentration in the presence of salt, have been explained by Rubinstein et al. [45] and Dobrynin et al. [46] with the help of dynamic scaling laws. In recent years, very extensive rheological data have become available on xanthan gum in aqueous solution in both the presence and absence of salt, and these have been analyzed in light of available theories [3, 47].

An unusual observation with polyelectrolytes is that, contrary to the behavior of nonionic polymers, the point of onset of shear thinning moves to higher shear rates as the polymer concentration in solution is increased. This is shown in Figure 4.8 for aqueous solutions of the sodium salt of sulfonated polystyrene [48]. The addition of salt makes the polymer size smaller and lowers the viscosity level, but the macromolecules become less sensitive to orientation and deformation, leading to a still-lower

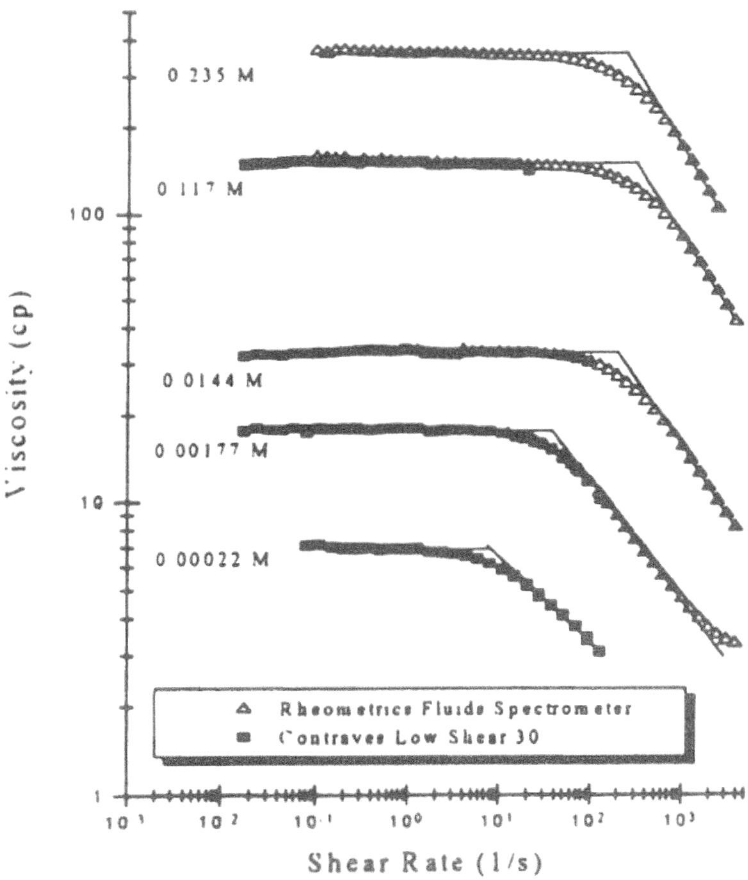

FIGURE 4.8 Shear rate dependence of the viscosity of aqueous solutions of the sodium salt of sulfonated polystyrene. Concentrations indicated are in moles of monomer per liter. (From Ref. [48].)

level of shear thinning [49]. These twin facts have been used to advantage by Tam and Tiu [50] to develop low- (but constant-) viscosity, highly elastic liquids by dissolving a few hundred ppm of polyacrylamide of 8–10 million molecular weight in saline solutions. The relaxation times of these ideal, elastic liquids are similar to those of polymer melts at processing temperatures, and this makes these constant-viscosity liquids valuable for experimentally simulating polymer melt behavior at room temperature.

The size of a charged polymer in solution can also be changed by changing the pH; this again influences both the intrinsic viscosity and the zero-shear viscosity. The latter quantity is shown in Figure 4.9 as a function of pH for aqueous solutions of a partially hydrolyzed polyacrylamide [51]; the hydrolysis of the nonionic polymer

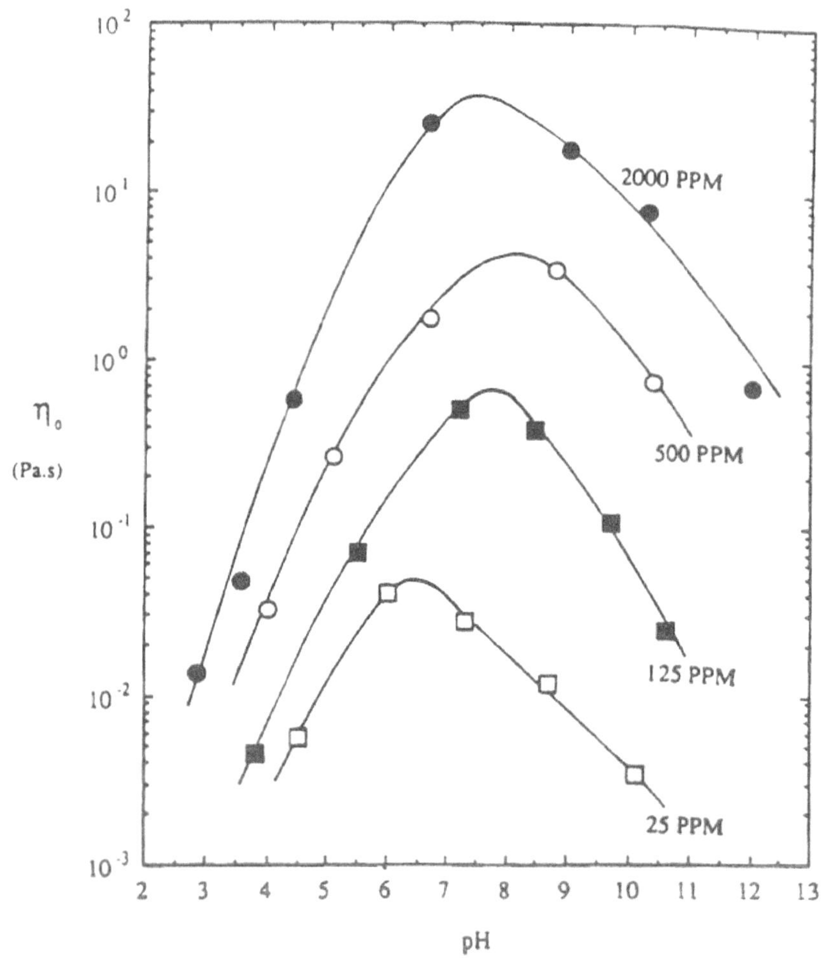

FIGURE 4.9 Effect of pH on the zero-shear viscosity of aqueous polyacrylamide solutions. (From Ref. [51].)

is generally effected with the help of an alkali such as NaOH or KOH. In solution, the pH level is changed by adding either 1 M HCl or 1 M NaOH. The maximum in the viscosity comes about because the polymer coil is most expanded at a slightly alkaline pH. At low pH values, the added hydrogen ions favor the formation of neutralized carboxyl groups on the polymer chain, making the polymer resemble a nonionic polymer that has a small coil size. At high pH values, the excess sodium ions are attracted to the carboxyl groups along the polymer chain, which again causes the polymer to coil up due to a local reduction in the negative charge on the chain.

VII. SURFACTANT SOLUTIONS AND ASSOCIATIVE THICKENERS

Synthetic detergents are surface-active agents that are small molecules with typical molecular weights ranging from 200 to 400. In very dilute solution, these *surfactants* are well dispersed, their behavior is Newtonian, and their zero-shear viscosity is only marginally greater than the viscosity of the solvent. They would ordinarily be of little interest to the rheologist, except that under appropriate conditions they can behave like polymer solutions and exhibit viscoelasticity, a non-Newtonian viscosity, and even a yield stress [52]. The presence of a yield stress is particularly advantageous in endowing stability against creaming or sedimentation to an emulsion prepared using the surfactant as an emulsifier. These non-Newtonian properties arise due to the ability of surfactant molecules to form large aggregates of different shapes.

Thus, as the concentration of surfactant molecules in solution is increased, a *critical micelle concentration* (CMC) is reached at which surfactant globules begin to be formed. At this stage, the solutions are still isotropic and of low viscosity. Upon increasing the concentration beyond the CMC, the globules grow into rodlike aggregates. The rodlike micelles, called *wormlike micelles*, can increase in length with further increases in surfactant concentration and become large enough to form entanglements with each other. If salt is added to this solution, a three-dimensional network or gel can form; this is shown schematically in Figure 4.10 [52].

A consequence of gelation is that the structural relaxation time can be as large as 10 sec, but it can be reduced significantly with the help of other additives without destroying the network structure. A mixture of cetylpyridinium chloride (CPyCl) and sodium salicylate (NaSal) in aqueous solution is an example of this progressive evolution of structures, and Figure 4.11 displays the zero-shear-rate viscosity of this system as a function of salt concentration [53]. The increase in viscosity in region II is due to the formation of entanglements between rodlike micelles. The other maxima and minima result from changes in the network relaxation time with changes in salt concentration. Usually, the shear viscosity can be written as a product of a shear modulus and a relaxation time. In the present instance, measurements show that it is the relaxation time that varies while the modulus remains unchanged as we go from region II to region V. Wormlike micelles find applications that are similar to polymer solutions, and these include use in fracturing fluids and as thickeners in personal care products. Their flow behavior can often be described by the Maxwell model with a single relaxation time [54]; this results from considering the micelles to be similar to a spring and dashpot in series (see Chapter 6).

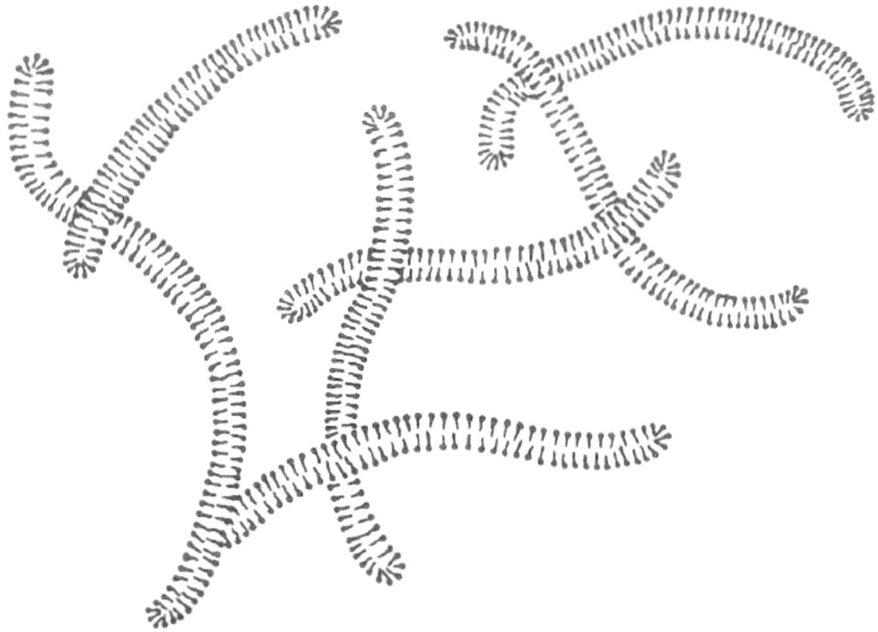

FIGURE 4.10 Schematic representation of a network of rodlike micelles in a surfactant solution. (From Ref. [52].)

Other *rheology modifiers* include thickeners like hydroxyethyl cellulose. When dissolved in water, these polymer molecules expand due to the formation of hydrogen bonds, and the solution viscosity increases. More versatile rheology modification is obtained with the use of *associative thickeners* that were developed due to the phaseout of organic solvents in paints and their replacement with water that is much more volatile. Associative thickeners are low-molecular-weight, water-soluble polymers that are similar to surfactants in that they increase the viscosity of water by associating with themselves and with any suspended solids or drops, providing desired flow performance. There are several types of associative thickeners that have achieved commercial acceptance, especially in the paints and coatings industry. Two of these are the hydrophobically modified ethylene urethane oxide rheology modifiers (*HEUR*) and the hydrophobically modified alkali swellable emulsions (*HASE*). The former type is the condensation product of the reaction between polyethylene glycol and a diisocyanate and is end-capped by hydrophobic groups such as long-chain alkanols. The typical molecular weight is about 40,000–50,000, and HEUR can be thought of as a double-ended surfactant molecule [55]. The chemical structure of a model HEUR molecule is shown in Figure 4.12 [56], and it has a hydrophilic middle but hydrophobic ends. While HEUR thickeners are nonionic, the HASE variety is anionic. Here, thickening takes place above a pH of 7 due to repulsion among carboxylate anions that are distributed along the polymer backbone. A major advantage of these thickeners is that the increase in viscosity is not accompanied by an increase in elasticity. Thus, paints that are thickened in this manner do not spatter as much as they would if cellulosic polymers had been employed.

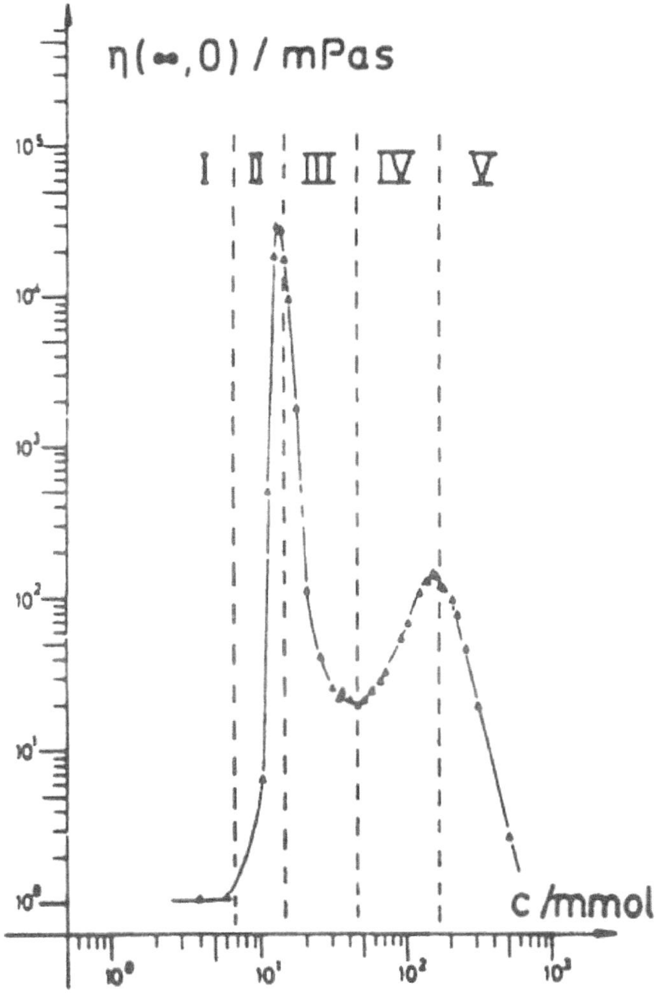

FIGURE 4.11 Zero-shear viscosity of a 15-mmol aqueous solution of CPyCl as a function of NaSal concentration. (Reprinted with permission from Ref. [53]. Copyright 1988 American Chemical Society.)

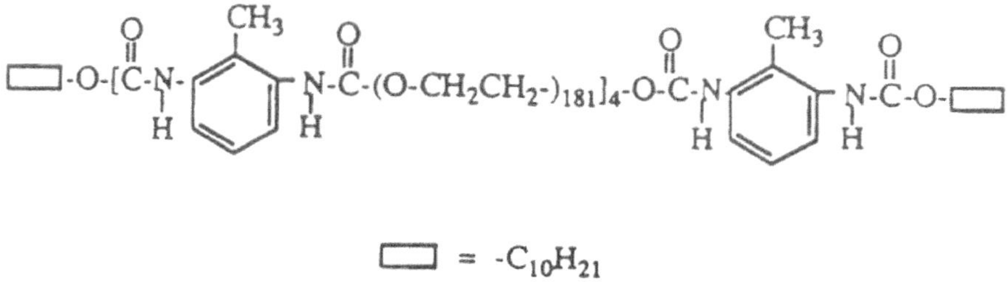

FIGURE 4.12 Structure of a model HEUR thickener. (From Ref. [56].)

In solution, HEUR molecules form a network by associating as micelles and also by adsorbing onto the surface of any suspended solids or drops. The resulting increase in viscosity can be used to retard the settling of suspended material. The network, however, is easily disrupted, and the thickening action at low shear rates depends on the surface area available for adsorption, while the viscosity at high shear rates is solely a function of polymer concentration. In the absence of suspended matter, intra- and intermolecular interactions make the zero-shear viscosity of HEUR thickeners increase rapidly with increasing concentration, more so than with unmodified polymers of similar molecular weight; the rate of viscosity increase becomes greater as the end-cap chain length is increased [57]. The viscosity generally remains Newtonian for an extended range of shear rates and then increases slightly and finally decreases. Sometimes thixotropy is observed. These and other observations have been explained by Annable et al. [57] with the help of transient-network models.

HASE-associative thickeners are modifications of traditional alkali-soluble thickeners that are carboxyl functional copolymers produced by free-radical polymerization of monomers such as methacrylic acid and ethyl acrylate [58]. The modification consists of the attachment of hydrophobic groups along the polymer backbone. These thickeners are essentially insoluble in water at low pH but dissolve at pH values above 7. Solution viscosity increases because of chain expansion due to repulsion between the carboxylate anions along the polymer backbone and also due to the formation of an intermolecular network resulting from hydrophobic interactions. The variation in zero-shear viscosity with pH and polymer concentration is shown in Figure 4.13 for aqueous solutions of an HASE-model polymer [58]. The solution viscosity is independent of polymer concentration and comparable to the solvent viscosity at low pH. However, as the pH increases beyond 5, the viscosity increases

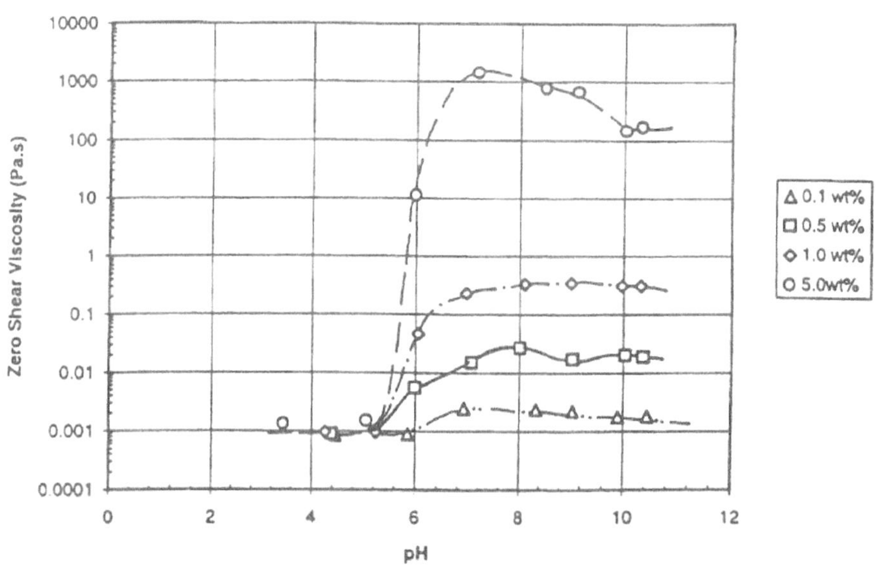

FIGURE 4.13 Zero-shear viscosity of HASE 5142 at different pHs. (From Ref. [58].)

suddenly, by as much as six orders of magnitude for the most concentrated solution. Thereafter, the change in viscosity with pH is quite gradual. A possible explanation for this behavior is provided is Figure 4.14. At low pH, the polymer exists as a dispersion rather than as a solution. When an alkali such as ammonium hydroxide is added to raise the pH, acid groups on the polymer chain are neutralized and the polymer dissolves. With increasing pH, the viscosity increases because (1) the polymer chain expands due to charge repulsion, and (2) intermolecular interactions lead to network formation. At high shear rates, these intermolecular contacts are disrupted, and substantial shear thinning is observed. The flow behavior of associative thickeners has been documented in a series of symposia edited by Glass [59–62].

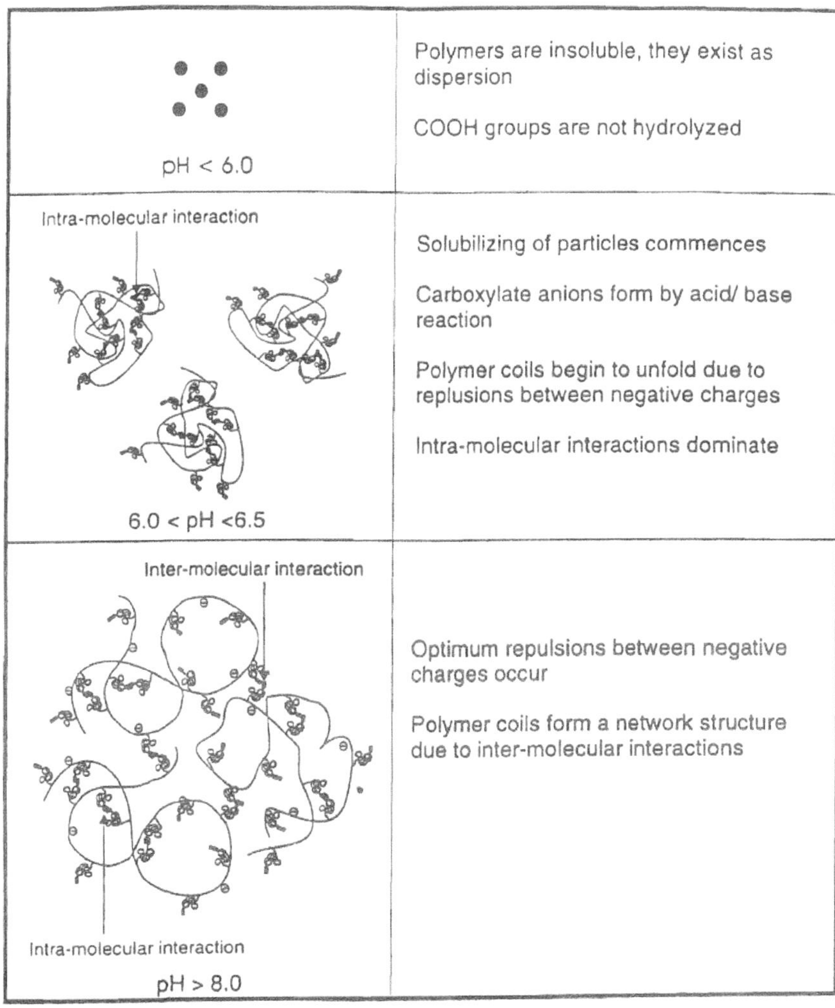

FIGURE 4.14 Pictorial representation of HASE polymer behavior with pH. (Unpublished results of Dr. K. C. Tam.)

The review by Larson et al. has a good discussion of associative thickeners with a focus on HEUR solutions [63]. This article provides an in-depth treatment of the effect of temperature as well as molecular weight and concentration of HEUR on the structure of the polymer in solution. Experimental data are presented on the different material functions, and these are explained based on available theories. The role of HEUR in thickening waterborne paints and modifying paint rheology for attaining desired flow behavior is also considered.

REFERENCES

1. G. O. Phillips and P. A. Williams. Handbook of Hydrocolloids. 3rd ed., Woodhead Publishing, Cambridge, MA, 2021.
2. W. W. Graessley, R. L. Hazleton, and L. R. Lindeman. The shear-rate dependence of viscosity in concentrated solutions of narrow-distribution polystyrene. Trans. Soc. Rheol. 11:267–285 (1967).
3. N. B. Wyatt and M. W. Liberatore. Rheology and viscosity scaling of the polyelectrolyte xanthan gum. J. Appl. Polym. Sci. 114:4076–4084 (2009).
4. D. V. Boger. A highly elastic constant-viscosity fluid. J. Non-Newt. Fluid Mech. 3:87–91 (1977/1978).
5. G. Prilutski, R. K. Gupta, T. Sridhar, and M. E. Ryan. Model viscoelastic liquids. J. Non-Newt. Fluid Mech. 12:233–241 (1983).
6. A. Einstein. Eine Neuve Bestimmung der Molekuldimension. Ann. Physik. 19:289–306 (1906).
7. A. Kumar and R. K. Gupta. Fundamentals of Polymer Engineering. 3rd ed., CRC Press, Boca Raton, FL, 2019.
8. P. J. Flory. Principles of Polymer Chemistry. Cornell University Press, Ithaca, NY, 1953.
9. M. Kurata and W. H. Stockmayer. Intrinsic viscosities and unperturbed dimensions of long chain molecules. Fortschr. Hochpolym.-Forsch. 3:196–312 (1963).
10. M. Kurata and Y. Tsunashima. Viscosity–molecular weight relationships and unperturbed dimensions of linear chain molecules. In: J. Brandrup and E. H. Immergut, eds. Polymer Handbook. 3rd ed., Wiley, New York, 1989, p. VII/1.
11. A. Lapasin and S. Pricl. Rheology of Industrial Polysaccharides: Theory and Applications. Springer, New York, 1995.
12. G. Marrucci and G. Ianniruberto. Molecular theories of polymer viscosity. In: A. Fasano, ed. Complex Flows in Industrial Processes. Springer, New York, 2000, pp. 3–24.
13. W. E. Rochefort and S. Middleman. Rheology of xanthan gum: Salt, temperature and strain effects in oscillatory and steady shear experiments. J. Rheol. 31:337–369 (1987).
14. E. M. Nsengiyumva, M. P. Heitz, and P. Alexandridis. Salt and temperature effects on xanthan gum polysaccharide in aqueous solutions. Int. J. Mol. Sci. 25:490 (2024).
15. R. K. Gupta. Polymer and Composite Rheology. 2nd ed., Marcel Dekker, New York, 2000.
16. A. M. Jamieson and R. Simha. Newtonian viscosity of dilute, semidilute, and concentrated polymer solutions. In: L. A. Utracki and A. M. Jamieson, eds. Polymer Physics: From suspensions to Nanocomposites and Beyond. Wiley, Hoboken, NJ, 2010, pp. 17–87.
17. D. R. Saini and A. V. Shenoy. Quick estimation of dilute polymer solution rheology and activation energy. J. Appl. Polym. Sci. 33:41–48 (1987).
18. T. E. Newlin, S. E. Lovell, P. R. Saunders, and J. D. Ferry. Long-range intermolecular coupling in concentrated poly-*n*-butyl methacrylate solutions and its dependence on temperature and concentration. J. Colloid Sci. 17:10–25 (1962).

19. H. Fujita and E. Maekawa. Viscosity behavior of the system polymethyl acrylate and diethyl phthalate over the complete range of composition. J. Phys. Chem. 66:1053–1058 (1962).
20. C. W. Macosko. Rheology. VCH, New York, 1994, p. 481.
21. P. F. Lyons and A. V. Tobolsky. Viscosity of polypropylene oxide solutions over the entire concentration range. Polym. Eng. Sci. 10:1–3 (1970).
22. K. S. Gandhi and M. C. Williams. Solvent effects on the viscosity of moderately concentrated polymer solutions. J. Polym. Sci.: Part C. 35:211–234 (1971).
23. F. Rodriguez. Suitability of equations for viscosity correlation over the entire concentration range. J. Polym. Sci.: Part B. 10:455–459 (1972).
24. F. Bueche. Viscosity, self-diffusion and allied effects in solid polymers. J. Chem. Phys. 20:1959–1964 (1952).
25. R. B. Bird, C. F. Curtiss, R. C. Armstrong, and O. Hassager. Dynamics of Polymeric Liquids. 2nd ed., Vol. 2. Wiley, New York, 1987.
26. P. E. Rouse Jr. A theory of the linear viscoelastic properties of dilute solutions of coiling polymers. J. Chem. Phys. 21:1272–1280 (1953).
27. W. W. Graessley. The entanglement concept in polymer rheology. Adv. Polym. Sci. 16:1–179 (1974).
28. E. R. Morris, A. N. Cutler, S. B. Ross-Murphy, and D. A. Rees. Concentration and shear rate dependence of viscosity in random coil polysaccharide solutions. Carb. Polym. 1:5–21 (1981).
29. Y. Takahashi, Y. Isono, I. Noda, and M. Nagasawa. Zero-shear viscosity of linear polymer solutions over a wide range of concentration. Macromolecules. 18:1002–1008 (1985).
30. F. N. Kelley and F. Bueche. Viscosity and glass temperature relations for polymer–diluent systems. J. Polym. Sci. 50:549–556 (1961).
31. W. W. Graessley. Effect of long branches on the flow properties of polymers. Accounts Chem. Res. 10:332–339 (1977).
32. L. E. Nielsen. Polymer Rheology. Marcel Dekker, New York, 1977.
33. J. D. Ferry. Viscoelastic Properties of Polymers. 3rd ed., Wiley, New York, 1980.
34. A. A. Tager and V. E. Dreval. Newtonian viscosity of concentrated solutions of polymers. Russian Chem. Rev. 36:361–373 (1967).
35. S. Yasutomi, S. Bair, and W. O. Winer. An application of a free volume model to lubricant rheology I—dependence of viscosity on temperature and pressure. J. Tribology. 106:291–303 (1984).
36. S. Bair and W. O. Winer. Some observations in high pressure rheology of lubricants. J. Lubrication Technol. 104:357–364 (1982).
37. C. S. Wu, E. E. Klaus, and J. L. Duda. Development of a method for the prediction of pressure-viscosity coefficients of lubricating oils based on free-volume theory. J. Tribology. 111:121–128 (1989).
38. K. C. Tam and C. Tiu. Steady and dynamic shear properties of aqueous polymer solutions. J. Rheol. 33:257–280 (1989).
39. M. Oritz, D. DeKee, and P. J. Carreau. Rheology of concentrated poly(ethylene oxide) solutions. J. Rheol. 38:519–539 (1994).
40. W. W. Graessley. Molecular entanglement theory of flow behavior in amorphous polymers. J. Chem. Phys. 43:2696–2703 (1965).
41. H. Morawetz. Macromolecules in Solution. 2nd ed., Wiley, New York, 1975.
42. M. Hara, ed. Polyelectrolytes. Marcel Dekker, New York, 1993.
43. R. M. Fuoss. Polyelectrolytes. Discuss. Faraday Soc. 11:125–134 (1951).
44. R. F. Fernandez Prini and A. E. Lagos. Tracer diffusion, electrical conductivity, and viscosity of aqueous solutions of polystyrenesulfonates. J. Polym. Sci.: Part A. 2:2917–2928 (1964).

45. M. Rubinstein, R. H. Colby, and A. V. Dobrynin. Dynamics of semidilute polyelectrolyte solutions. Phys. Rev. Lett. 73:2776–2779 (1994).
46. A. V. Dobrynin, R. H. Colby, and M. Rubinstein. Scaling theory of polyelectrolyte solutions. Macromolecules 28:1859–1871 (1995).
47. E. M. Nsengiyumva and P. Alexandridis. Xanthan gum in aqueous solutions: Fundamentals and applications. Int. J. Biol. Macromol. 216:583–604 (2022).
48. D. C. Boris and R. H. Colby. Shear viscosity of polyelectrolyte solutions. Proc. XIIth Int. Congr. Rheol., Quebec City, Canada, 1996, pp. 195–196.
49. A. Ait-Kadi, P. J. Carreau, and G. Chauveteau. Rheological properties of partially hydrolyzed polyacrylamide solutions. J. Rheol. 31:537–561 (1987).
50. K. C. Tam and C. Tiu. A low viscosity, highly elastic ideal fluid. J. Non-Newt. Fluid Mech. 31:163–177 (1989).
51. K. C. Tam and C. Tiu. Effect of pH on the excluded volume of high molecular weight anionic polyacrylamide. Polym. Commun. 31:102–104 (1990).
52. H. Hoffmann and G. Ebert. Surfactants, micelles and fascinating phenomena. Angew. Chem. Int. Ed. Engl. 27:902–912 (1988).
53. H. Rehage and H. Hoffmann. Rheological properties of viscoelastic surfactant systems. J. Phys. Chem. 92:4712–4719 (1988).
54. M. In. Linear rheology of aqueous solutions of wormlike micelles. In: R. Zana and E. W. Kaler, eds. Giant Micelles: Properties and Applications. CRC Press, Boca Raton, FL, 2007, pp. 249–287.
55. J. Prideaux. Rheology modifiers and thickeners in aqueous paints. Surf. Coat. Int. J. Oil Color Chemists' Assoc. 76:177–182 (1993).
56. D. L. Lundberg, J. E. Glass, and R. R. Eley. Viscoelastic behavior among HEUR thickeners. J. Rheol. 35:1255–1274 (1991).
57. T. Annable, R. Buscall, R. Ettelaie, and D. Whittlestone. The rheology of solutions of associating polymers: comparison of experimental behavior with transient network theory. J. Rheol. 37:695–726 (1993).
58. K. C. Tam, M. L. Farmer, R. D. Jenkins, and D. R. Bassett. Rheological properties of hydrophobically modified alkali-soluble polymers: Effects of ethylene-oxide chain length. J. Polym. Sci. B: Polym. Phys. 36:2275–2290 (1998).
59. J. E. Glass, ed. Water-Soluble Polymers: Beauty with Performance. Advances in Chemistry Series 213. American Chemical Society, Washington, DC, 1986.
60. J. E. Glass, ed. Polymers in Aqueous Media: Performance Through Association. Advances in Chemistry Series 223. American Chemical Society, Washington, DC, 1989.
61. J. E. Glass and D. N. Schulz, eds. Polymers as Rheology Modifiers. ACS Symposium Series 462. American Chemical Society, Washington, DC, 1991.
62. J. E. Glass, ed. Hydrophilic Polymers: Performance with Environmental Acceptability. Advances in Chemistry Series 248. American Chemical Society, Washington, DC, 1996.
63. R. G. Larson, A. K. Van Dyk, T. Chatterjee, and V. V. Ginzburg. Associative thickeners for waterborne paints: Structure, characterization, rheology, and modeling. Prog. Polym. Sci. 129:101546 (2022).

5 Normal Stress Differences in Polymers During Shear Flow

I. INTRODUCTION

Normal stresses were defined in Chapter 1. Unequal normal stresses in shear flow are not found with Newtonian liquids but are characteristic of non-Newtonian polymer melts and solutions. Normal stress differences are primarily manifestations of the elasticity of polymeric materials, and this elasticity results from the orientation and extension of polymer chain segments. Chain extension and orientation are aided by the presence of entanglements, and this results in the storage of energy within the polymeric fluid.

If we consider steady-shear flow in a Cartesian coordinate system, it is reasonable to expect that the typical polymer molecule will stretch due to the presence of a velocity gradient and will also tend to align itself with the flow field; this will result in the presence of a force f along the polymer chain axis, as shown in Figure 5.1. If the polymer molecule lies in the x_1–x_2 plane, the angle that the force vector makes with the x_1 axis will be less than 45° and will decrease progressively with increasing shear rate. Thus, for flexible macromolecules, the component of the external force f along x_1 will always be greater than the corresponding component along x_2. As a result, the polymer contribution to the normal stress T_{11} will be larger than the contribution to T_{22}, and this will make the first normal stress difference, N_1, a positive quantity. The second normal stress difference, N_2, was once thought to be zero for many years, but it is now accepted to be nonzero and negative but significantly smaller than N_1 in magnitude [1, 2]. The first and second normal stress coefficients are defined as $\psi_1 = \frac{N_1}{\dot{\gamma}^2}$ and $\psi_2 = \frac{N_2}{\dot{\gamma}^2}$, respectively. At very low shear rates, both normal stress differences increase with the square of the rate of shear, so the corresponding normal stress coefficients become constant in the limit of vanishingly low deformation rates, indicating linear viscoelastic behavior. The low shear rate limiting value of the first normal stress coefficient can be obtained from dynamic mechanical experiments as follows [3]:

$$\psi_{10} = \lim_{\dot{\gamma}\to 0} \psi_1 = 2 \lim_{\omega\to 0} \frac{G'}{\omega^2} \tag{5.1}$$

Equation (5.1) is a useful result, since it is often easier to measure the storage modulus as compared to the first normal stress difference. However, it should be

DOI: 10.1201/9781003013501-5

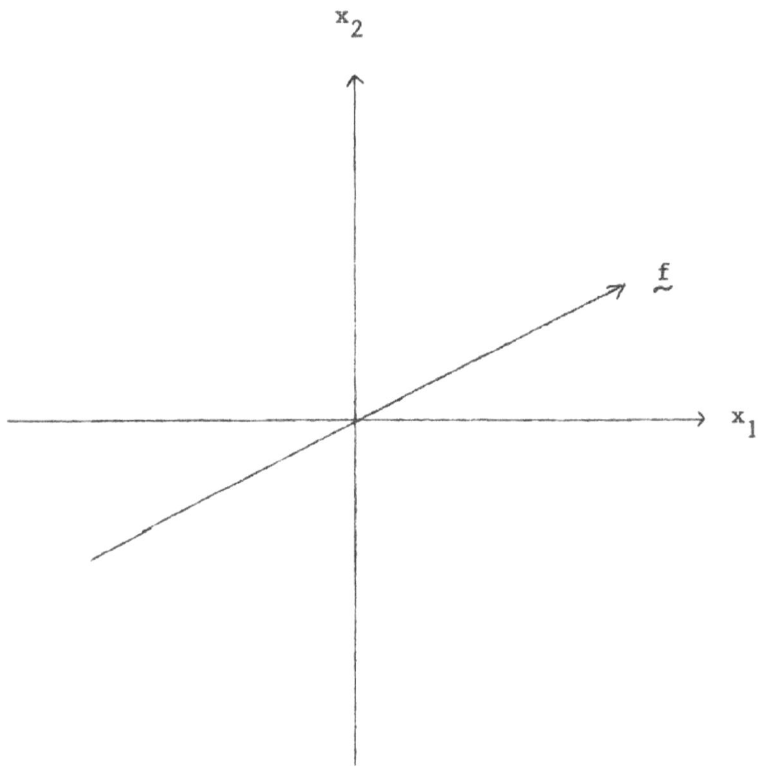

FIGURE 5.1 Force acting on a typical polymer molecule during shear flow. (From Ref. [2].)

recognized that polymer molecules are only slightly perturbed around their equilibrium positions during the measurement of dynamic moduli. By contrast, normal stress difference measurements involve large strains which are more representative of polymer processing operations. In general, the normal stress coefficients decrease with increasing shear rate; however, the high shear rate limit is uncertain, owing to the difficulty of making accurate normal stress measurements at high shear rates.

Figure 5.2 shows representative first normal stress difference data for polymer melts as a function of shear rate at three constant temperatures [4]. These data are for a commercial sample of polystyrene and were obtained on a Weissenberg rheogoniometer employing cone-and-plate fixtures. Clearly, stress levels decrease on increasing the temperature of measurement. Also, on logarithmic coordinates, the slope of each of the three curves increases as the shear rate decreases. However, data do not extend to low-enough shear rates for us to observe the expected quadratic dependence of N_1, on the shear rate. Most of the early work on polymer solutions was carried out on concentrated solutions of polyisobutylene, and Figure 5.3 shows the steady-state shear stress and both normal stress differences for this polymer dissolved in decalin [5]. Measurements were made at room temperature and utilized cone-and-plate as well as parallel-plate fixtures. The variation of the first normal

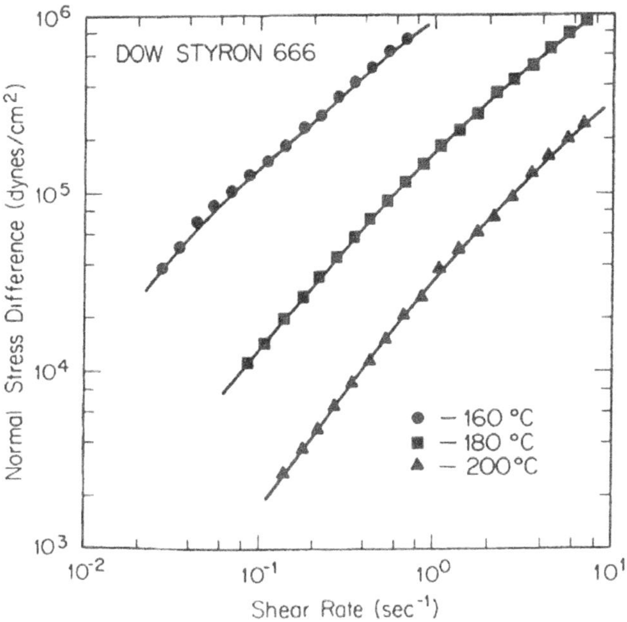

FIGURE 5.2 First normal stress difference as a function of shear rate and temperature for molten polystyrene. (From Ref. [4].)

stress difference with shear rate is qualitatively similar to the behavior of molten polymers. Note that the first normal stress difference curve lies below the shear stress curve at low shear rates, but this trend is reversed upon increasing the shear rate. This happens because of the stronger dependence of N_1, on the shear rate compared with the dependence of τ on $\dot{\gamma}$. The ratio of N_1 to the shear stress is often used as a measure of fluid elasticity; this ratio tends to zero as the shear rate tends to zero. Values of the second normal stress difference are not as accurate as those of the first normal stress difference, but these are clearly of opposite sign and are almost an order of magnitude smaller. Stress levels as well as the precision of measurement decrease as the polymer concentration in solution is decreased [5]. Since both normal stress coefficients are even functions of the shear rate and since they become constant at small shear rates, the variation of either of them with shear rate can be represented mathematically as follows:

$$\psi_i = \frac{a_i}{1 + b_i \dot{\gamma}^2} \quad i = 1, 2 \tag{5.2}$$

where a_i and b_i are constants. This form also emerges from most rheological models, whether fundamental or empirical and whether based on network theory or dilute solution theory [6, 7].

Upon inception of flow at a constant shear rate, steady-state stress levels are neither attained instantly nor reached monotonically. Unless the shear rate is extremely

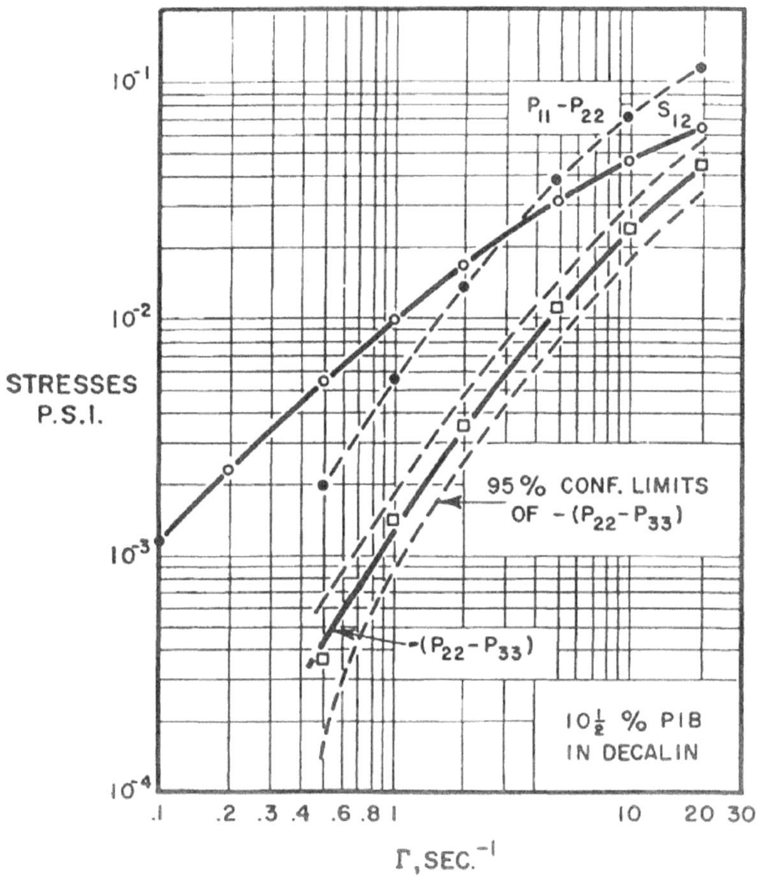

FIGURE 5.3 Stress levels, steady laminar shearing flows, for 10.5% polyisobutylene in decalin. Note that P_{11}–P_{22} is N_1, and P_{22}–P_{33} is N_2, and the x-axis is shear rate. (From Ref. [5].)

low, the shear stress as well as the normal stresses exhibit overshoots relative to their steady-state values, and there may be more than one oscillation [8, 9]. The percentage by which the various material functions overshoot their steady-state values increases with increasing shear rate, and several minutes may be needed for the transients to die out. Peak N_1 values observed by Meissner for a polyethylene melt were as much as four times the steady-state value [8]. The shear stress transients are similar to the normal stress transients, but the maximum in the shear stress is relatively smaller than the maximum in the first normal stress difference and occurs before it [8]. For molten polyethylene, the shear stress maximum is observed at a total strain of 7; the normal stress maximum occurs at a total strain of 14 [10]. For polybutadiene solutions, Menezes and Graessley [11] found that while the ratio of the strain values at which N_1 and the shear stress were a maximum was approximately 2.5, the strain at the overshoot peak in N_1 was 5 at low shear rates and increased progressively with increasing shear rate. Transient effects arise on cessation of shearing as well. Stresses decay monotonically to zero, with the shear stress relaxing faster than the normal stress differences; the rate of decay depends on the shear rate prior to cessation of shearing [9].

Normal stresses produce a number of phenomena not found with Newtonian liquids. For example, when a polymer is extruded from an orifice, a capillary, or a slit, the diameter or thickness of the resulting strand is considerably greater than the diameter of the hole from which it came. This phenomenon, shown in Figure 5.4, is called *die swell* or *jet swell* or *extrudate swell.* That die swell is caused by the presence of N_1 resulting from fluid being sheared in the capillary is easy to explain. As liquid emerges into the atmosphere, the total stress in the flow direction is the negative of the atmospheric pressure. A positive N_1 implies that the total stress in the radial direction is negative (compressive) and, in magnitude, is greater than the pressure of the atmosphere. In other words, the tube wall pushes down on the liquid being sheared, and the liquid pushes up on the tube with an equal stress that exceeds

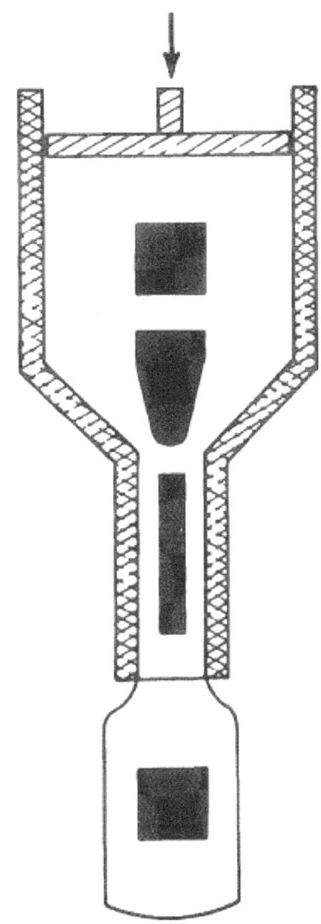

FIGURE 5.4 Schematic diagram illustrating the phenomenon of die swell. (From Ref. [12].)

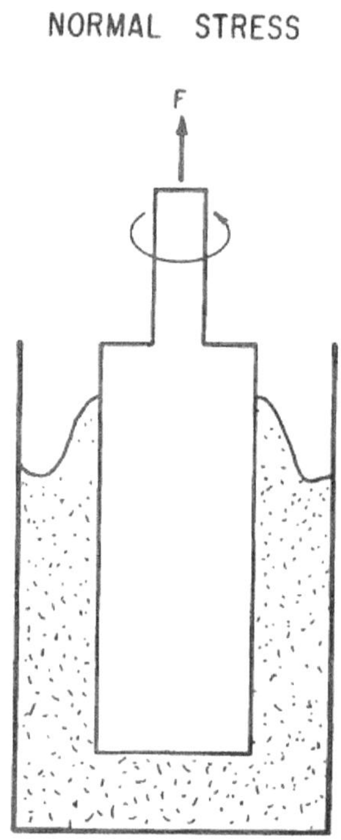

FIGURE 5.5 Creep of a polymeric liquid up the rotating inner cylinder of a coaxial cylinder rheometer. (From Ref. [12].)

the atmospheric pressure in magnitude. When the fluid emerges into the atmosphere, there is no constraining tube; the fluid snaps back like a stretched rubber band, and we observe die swell.

Another unexpected phenomenon due to normal stresses is the *Weissenberg effect*, or the ascending of polymer liquids up rotating shafts, as illustrated in Figure 5.5. In contrast, rotating shafts in Newtonian liquids cause a depression of the liquid surface because of centrifugal forces. In cone-and-plate rheometers, and in other rotating systems of similar geometry, the normal stresses tend to force the cone-and-plate apart; a measurement of this force is the most reliable method of obtaining the first normal stress difference, and this is the principle by which N_1 is determined in the commercial instruments described in Chapter 2. If holes are drilled in the plates parallel to the rotating axis, liquid will be forced up through the holes. This is the basis of one early technique for calculating normal stresses by measuring the height to which a liquid will ascend up a tube for a given speed of rotation [13, 14]. The first normal stress difference is proportional to the height of the liquid in the tube, and the liquid level is a maximum for a tube at the center of rotation and decreases toward the edge of the rotating disk. Instead of measuring liquid levels, we may use pressure transducers located along the radius of the plate [15]. At low shear rates, the normal stresses are proportional to the square of the shearing stress, so that a doubling of the speed of rotation will increase the normal stress by a factor of nearly 4.

Normal stresses can be important in a number of polymer processing situations. In wire coating, normal stresses help produce a smooth coating of uniform thickness up to the point where melt fracture instability takes over, and at the same time they keep the wire properly centered if the second normal stress difference is negative [16]. If a polymer melt is sheared between a rotating plate and a stationary plate, the tendency of the fluid to push up against the plates under the influence of a positive first normal stress difference can be utilized to extrude the melt continuously. This can be done by punching a hole at the center of the stationary plate, as shown in Figure 5.6. This screwless, normal stress extruder was originally proposed by Maxwell and Scalora [17], and it was commercially available from Custom Scientific Instruments in Cedar Knolls, NJ. A major advantage of the Maxwell extruder is that only gram quantities of polymer are required, and this is helpful when working with experimental materials. Even though the residence time is short, polymer blending can be carried out and shear-sensitive as well as hygroscopic polymers accommodated.

The different methods of measuring the first normal stress difference in shear are considered next. Note, though, that all these methods also allow for simultaneous measurement of the shear viscosity. However, N_1 turns out to be much more sensitive to small changes in material characteristics than does the viscosity, and this is a major motivating factor for making normal stress difference measurements [18].

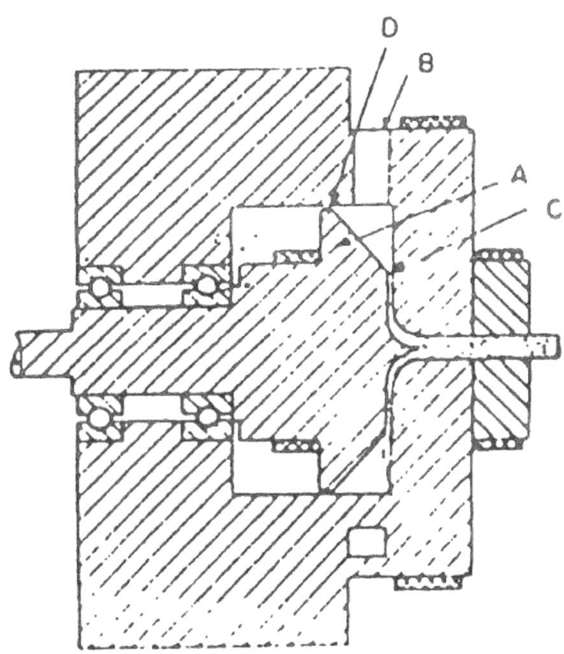

FIGURE 5.6 Normal stress extruder. A is the rotor, polymer is fed at point B, and the gap between the plates is given by the spacing at point C. (From Ref. [17]. Reprinted with permission from *Modern Plastics*, a publication of Chemical Week Associates, New York.)

II. LOW-SHEAR-RATE BEHAVIOR OF THE FIRST NORMAL STRESS DIFFERENCE

The commercial availability of very sophisticated rotational viscometers has made the determination of the first normal stress difference a straightforward procedure for polymer melts, polymer solutions, and certain food dispersions (see, for example, [19]). The normal force range over which measurements can be made is fairly broad, and, in the TA Instruments AR 1000-N instrument, for example, forces can be measured reliably from 0.01 N to 50 N. In this respect, cone-and-plate geometry is the preferred one, but, as mentioned in Chapter 2, secondary flows and instabilities limit the shear rate range of measurement to a maximum of a few hundred reciprocal seconds. While these deformation rates are significantly smaller than those often encountered in industrial operations, enough data are available to allow us to make definite statements concerning the influence of variables such as temperature, shear rate, concentration, and molecular structure on the measured values of N_1.

As is evident from Figures 5.2 and 5.3, the first normal stress difference of a polymer melt or a polymer solution at a fixed concentration depends on both temperature and shear rate. It is a remarkable fact, however, that all the data for a particular fluid collapse onto a single straight line when the logarithm of N_1 is plotted against the logarithm of the shear stress. This is shown in Figure 5.7 for the melt data of Figure 5.2 and in Figure 5.8 for data on liquid M1 [20], a 0.244% solution of high-molecular-weight polyisobutylene in a mixed solvent of kerosene and low-molecular-weight

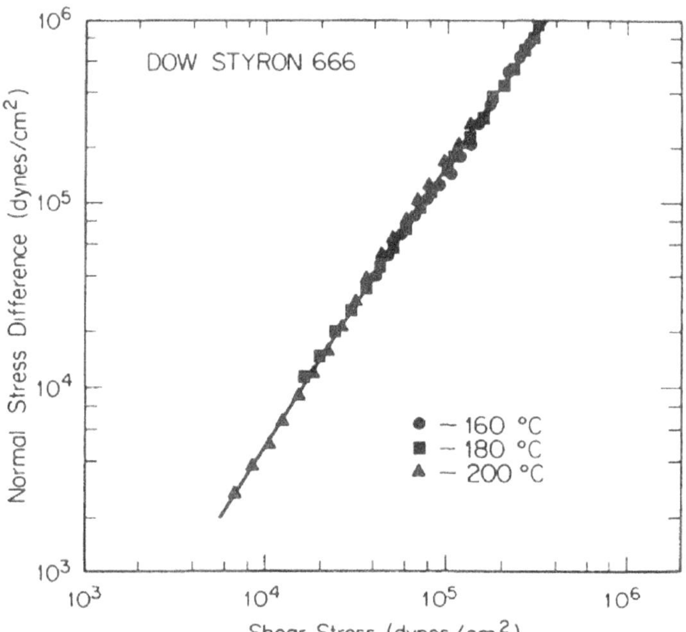

FIGURE 5.7 First normal stress difference as a function of shear stress for the data of Figure 5.2. (From Ref. [4].)

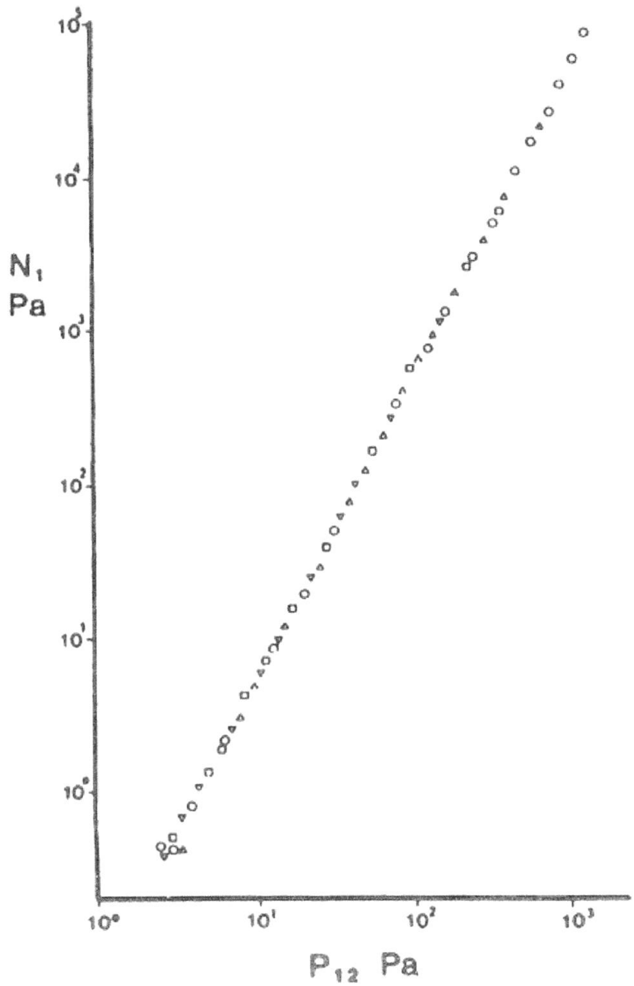

FIGURE 5.8 First normal stress difference as a function of shear stress for fluid M1. The different symbols are for different temperatures that vary from 20 to 50°C in intervals of 10°C. (From Ref. [20], with permission from Elsevier Science.)

polybutene [21]. The slope of the line in Figure 5.7 is 1.66, and that of the line in Figure 5.8 is 2. In general, we can say that

$$N_1 = A\tau^a \tag{5.3}$$

in which A and a are constants, and τ is the shear stress. Thus, a = 1.66 for polystyrene. The maximum observed value of the slope a, and this corresponds to behavior in the linear viscoelastic range, is 2. Another remarkable feature of first normal stress difference data when plotted versus shear stress rather than against shear rate is the narrowing of differences among different polymers [22] or among different molecular weight fractions of the same polymer [23]. The utility of Eq. (5.3), though, goes beyond data correlation alone. Since it is relatively easy to compute or to measure the shear stress, one often uses Eq. (5.3) to determine the first normal stress difference, even in polymer processing situations [24]. Equation (5.3) also tells us that N_1 can easily exceed τ by an order of magnitude under normal polymer melt processing conditions. Since N_1 is more sensitive than τ is to the high-molecular-weight tail of the polymer molecular weight distribution, a measurement of N_1 can suggest likely differences in the behavior of apparently similar polymers during processing that cannot be revealed by a measurement of the shear stress alone. Note that the magnitude of N_1 increases as the molecular weight distribution broadens [25]. Furthermore, the ratio of the first normal stress difference in shear to twice the corresponding shear stress is a dimensionless group designated as S_R and known as the *recoverable shear*. A critical value of this group is frequently employed to predict the onset of elastic flow instabilities, such as melt fracture [26]; a typical value of S_R at the onset of melt fracture is 10 [25]. Note that the ratio of the recoverable shear to the shear rate is considered to be a characteristic fluid time, because it has the dimensions of time and a value that tends to a constant as the shear rate tends to zero. This characteristic time, also called the *fluid relaxation time*, is a measure of fluid elasticity, and it represents the time duration over which stresses persist after cessation of shearing; this interpretation follows from the predictions of some of the most common rheological models. On dividing the relaxation time by the duration time of a process of interest, one gets a dimensionless number called the *Deborah number*. Fluid elasticity is important only for those processes for which the Deborah number is non-negligible [27].

If the dependence of shear stress on variables such as temperature and shear rate is known, the dependence of the first normal stress difference in shear on the same variables follows directly from Eq. (5.3). In particular, N_1 decreases with increasing temperature, and the polymer becomes less elastic as well, because the relaxation time goes down owing to the fact that the exponent a in Eq. (5.3) exceeds unity. Additionally, Eq. (5.3) predicts that time–temperature superposition must hold for the first normal stress difference when plotted as a function of the shear rate, just as it does for the shear stress (see Figures 3.14 and 3.15); the temperature shift factors in the two cases are identical. Since the melt viscosity of monodisperse, linear polymers varies with the 3.4 or 3.5 power of the molecular weight at low shear rates (see Section II.A in Chapter 3), we can expect the zero-shear-rate first normal stress difference or the first normal stress coefficient to increase with molecular weight to about the power 7; nylon 6 melt data of Laun [28], shown in Figure 5.9, fulfill this expectation. Regarding the influence of molecular weight distribution, it appears that broadening the distribution of a linear polymer, while keeping the average molecular weight and the shear stress unchanged, results in an increase in the normal stress difference but a decrease in the exponent a in Eq. (5.3) [4, 29]. According to Minoshima et al. [23], data on molten polystyrenes suggest that the constant A in Eq. (5.3) may be written as

$$A = A'\left(\frac{M_z}{M_w}\right)^{3.4} \tag{5.4}$$

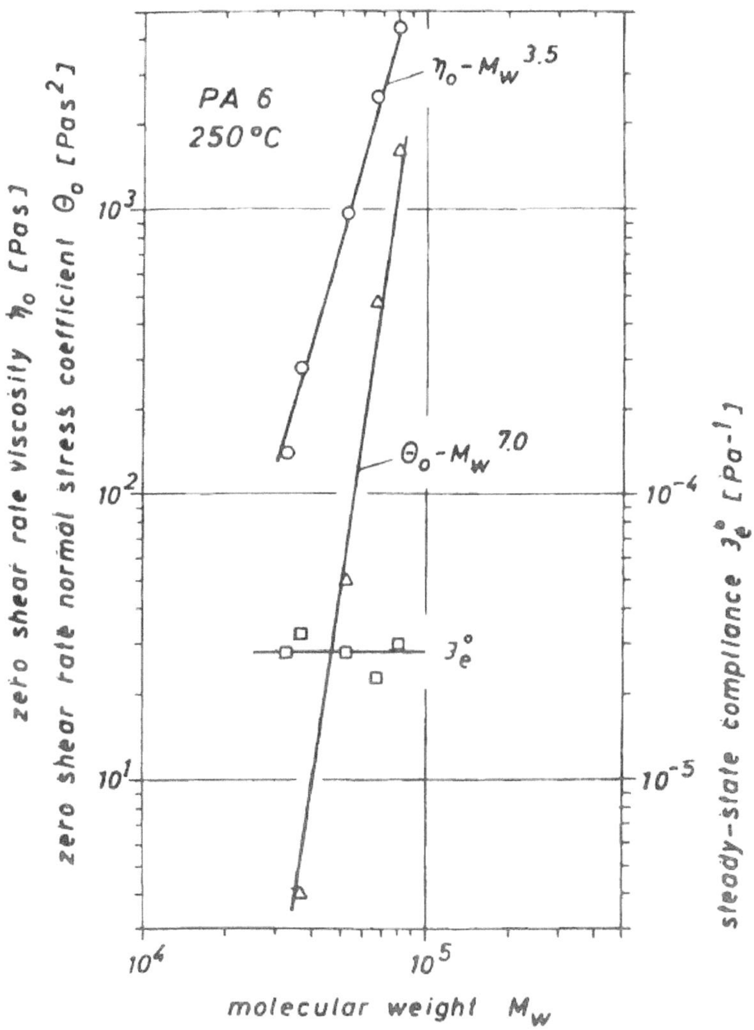

FIGURE 5.9 Zero-shear viscosity and the first normal stress coefficient of a nylon 6 melt as a function of molecular weight. (From Ref. [28].)

where A' is another constant and M_z is the z-average molecular weight [30]. The effect of the introduction of long-chain branching is unclear. While the polyethylene data of Mendelson et al. [31] show a decrease in the first normal stress difference, the reverse is found to be true for branched polyethylene terephthalate [32].

The preceding remarks have, to a large extent, been made with reference to the behavior of polymer melts. However, the shear rate, temperature, molecular weight, and molecular weight distribution dependence of the first normal stress difference of polymer solutions is similar to that of melts [33]. Now, though, polymer concentration and solvent viscosity are additional variables; increasing the value of either

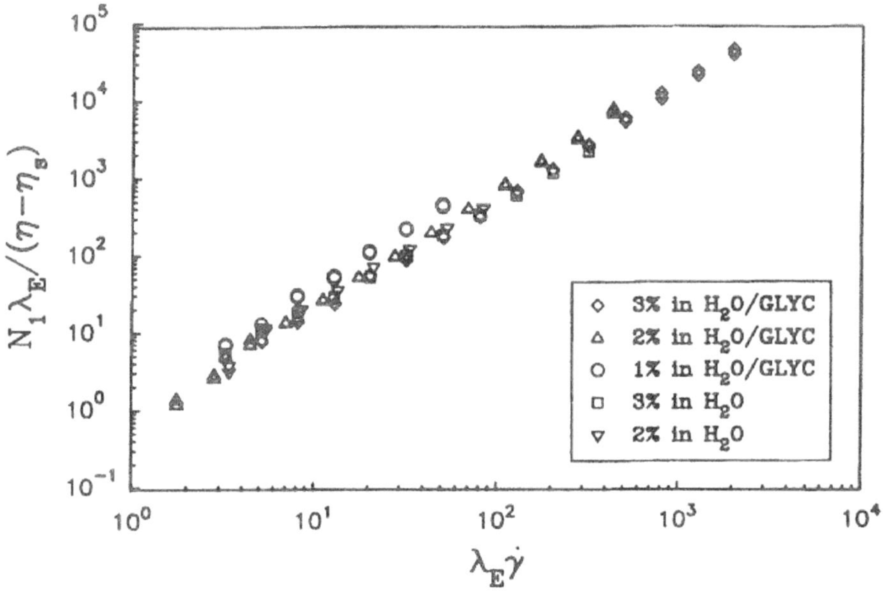

FIGURE 5.10 Dimensionless first normal stress difference versus dimensionless shear rate for PEO solutions in two different solvents. The average molecular weight is 1.8 million. (From Ref. [35].)

quantity results in an increase in N_1. Low-shear-rate data on polyisobutylene solutions [34] suggest that the group N_1c^2/τ^2 is essentially constant. At higher shear rates, data can be represented as dimensionless plots on logarithmic coordinates. The result is a master curve that is independent of polymer concentration, polymer molecular weight, and solvent viscosity. Figure 5.10 shows such a plot of the dimensionless first normal stress difference of poly (ethylene oxide) solutions as a function of the dimensionless shear rate taken from the work of Oritz et al. [35]. N_1 is made dimensionless by multiplying it with $\lambda_\varepsilon/(\eta-\eta_s)$, while the shear rate is made dimensionless by multiplying it with λ_ε. Here η is the solution viscosity, η_s is the solvent viscosity, and λ_ε is the same time constant defined earlier by Eq. (4.23). Note that the normal stress data on polymer melts can also be plotted in the form of temperature-and-molecular-weight-independent master curves; such unified curves are useful for estimating the normal stress response of chemically similar polymers [36].

In order to explain the observed relationship between the first normal stress difference function of polymeric fluids and the various material and processing variables, we can take one of two approaches. In the phenomenological approach, we can relate the three-dimensional stress to the history of deformation in a general way and then examine limiting cases. For *simple fluids* with fading memory, the first order approximation for slow flows gives the Newtonian fluid [37]. The next-higher approximation, known as the *second-order fluid*, possesses viscoelasticity. For this fluid, the normal stress differences are nonzero; in particular, the first normal stress difference is predicted to increase quadratically with shear rate. As we have seen, this behavior is observed at low shear rates, but transient effects, such as stress relaxation, cannot be represented using the second-order fluid model. However, stress relaxation, stress overshoots, and a less-than-quadratic increase in the first normal stress difference at high shear rates can be explained with the help of more general constitutive equations derived using a continuum mechanics approach [9, 37]. These equations, though, suffer from the handicap that model parameters cannot be related to polymer structural variables; to accomplish this, we have to employ a molecular approach. Three different molecular models are popular with rheologists: *bead–spring models* for dilute solutions, *network models* for melts, and reptation models for concentrated solutions and melts [38].

The essential elements of bead–spring theories were mentioned in Section III of Chapter 4. In the simplest situation, each polymer molecule is modeled as a dumbbell that consists of two equal masses connected by an infinitely extensible, linear, elastic spring [39, 40]. The use of this model results in a closed-form, two-constant, constitutive equation for the polymer contribution to the stress in the solution. This equation is called the *upper convected Maxwell equation*. It captures all the main qualitative features of polymer viscoelasticity, and for this reason it has been used extensively to model polymer processing operations [41]. The Maxwell equation predicts that the first normal stress difference is given by

$$N_1 = \frac{2\theta\tau^2}{\eta} \tag{5.5}$$

in which θ and η are, respectively, the constant-relaxation time and shear viscosity. Thus, Eq. (5.5) is the same as Eq. (5.3) if a is taken to be 2. However, the constants in Eq. (5.5) have a physical meaning. In particular, the relaxation time can be related to molecular variables as follows [42]:

$$\theta = \frac{K\eta_s \mathrm{M}^{3/2}\alpha^3}{T} \tag{5.6}$$

where K is a constant, and α is a coil expansion factor whose value depends on solvent quality and temperature. Clearly, the combination of Eqs. (5.5) and (5.6) reveals the influence of shear rate, temperature, molecular weight, solvent viscosity, and solvent quality on the first normal stress difference in shear. These equations have been verified for dilute polymer solutions up to moderate shear rates, and Tam et al. [42, 43] have used this theoretical framework to formulate a large number of constant-viscosity, ideal elastic liquids; typical data are displayed in Figure 5.8. These liquids, which can have relaxation times comparable to those of polymer melts at processing temperatures, are known as *Boger fluids* [44] after D. V. Boger, who first formulated them; their constitutive behavior was first explained by Prilutski et al. [45]. At high shear rates, the first normal stress difference of dilute polymer solutions deviates from the predictions of Eq. (5.5). As a consequence, this equation has been modified by changing some of the model assumptions, such as by making the dumbbells finitely extensible and the spring law nonlinear [46].

Network models employed for representing the flow behavior of polymer melts owe their origin to the theory of rubber elasticity. Unlike vulcanized rubber, the network junctions are temporary rather than permanent crosslinks. It is remarkable, however, that the simplest constitutive equation that emerges from the use of network theory is again the upper convected Maxwell equation (also known as the *Lodge rubberlike liquid* in the case of polymer melts). To accommodate values of the exponent a in Eq. (5.3) that become less than 2 with increasing shear rate, different assumptions can be made concerning the rates of creation and destruction of network points. This has been an area of active research, and a very large number of network-type models have been proposed in the literature [38]. The introduction of reptation models was an attempt to describe better the dynamics of concentrated polymer solutions and polymer melts. Reptation models were mentioned in Chapter 3, and the reader is referred to the technical literature for details [47, 48].

III. MEASUREMENT OF N_1 AT HIGH SHEAR RATES

Due to problems of edge fracture and sample loss, the upper limit on shear rate for a cone-and-plate viscometer is about 10 sec^{-1} for polymer melts and a few hundred reciprocal seconds for polymer solutions. To obtain data at higher deformation rates, one uses capillary and slit devices to infer the first normal stress difference from a measurement of quantities such as the exit pressure, the jet thrust, die swell, and the hole pressure. To understand these quantities, consider Figure 5.11 which shows flow

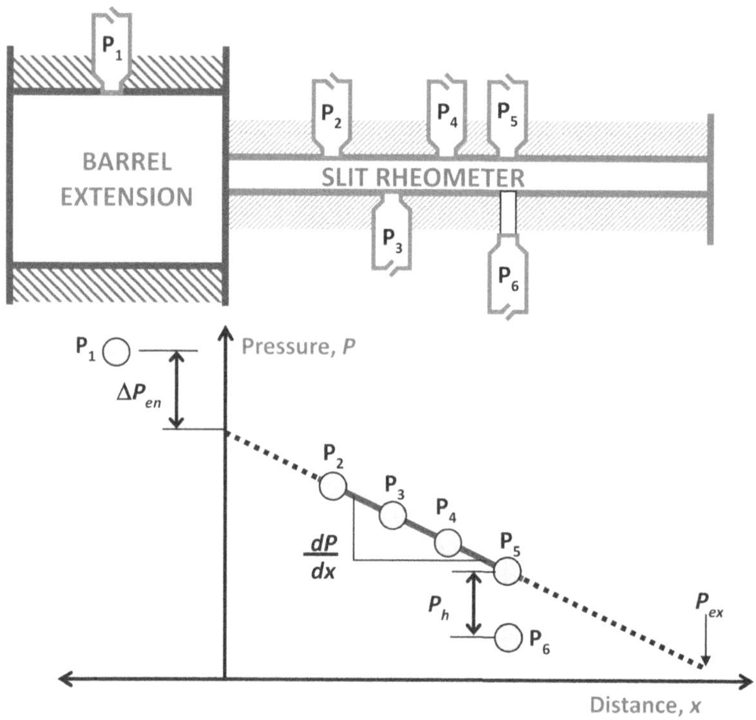

FIGURE 5.11 Possible pressure measurements during flow through a slit rheometer. (Adapted from Ref. [18].)

through a slit die. The die length is such that fully developed flow can be established within it. The flow itself is generated with the help of an extruder fitted with a barrel extension; alternatively, a capillary rheometer may be used to generate the flow. Steady-state pressure values are measured within the barrel extension and at different places along the length of the slit die, as shown in Figure 5.11.

The flow in the region between pressure transducers P_2 and P_5 is fully developed, and, as a consequence, there is a linear decrease in pressure as one goes from point P_2 to P_5. However, if the pressure readings are extrapolated backward to x = 0, the extrapolated value is significantly lower than the pressure measured by transducer P_1. The difference between P_1 and the extrapolated entry pressure is known as the *entry pressure* loss, Δp_{en}. It results from non-viscometric flow near the die entrance, and it is found that the flow field has an extensional component. As shown in Chapter 7, Δp_{en} can be employed to estimate the extensional viscosity of the fluid. The flow is non-viscometric near the die exit as well. The extrapolated pressure at that location exceeds atmospheric pressure, and the difference between the two values is called the *exit pressure* P_e. This is small in magnitude, and its variation with the wall shear stress can be used to determine N_1. Additionally, if a pressure transducer P_6 is located in a recessed region directly across from P_5, $(P_5–P_6)$ is found to be positive, and it is termed the *hole pressure*. The magnitude of the hole pressure is comparable to the magnitude of the exit pressure, and its variation with the wall shear stress can similarly be used to determine N_1. Finally, the thickness of the sheet leaving the die exceeds the die gap, and this too can be employed for N_1 measurement. Although these techniques are firmly established in the rheological literature and commercial instruments were available at different times in the past, the theory relating N_1 to any of the measured quantities involves assumptions that may not be strictly valid in all cases. One, therefore, has to exercise some caution in data interpretation. We also mention that in a new commercial instrument, a radial flow channel is added to the end of the slit, and theory is developed to calculate N_1 from the measurement of pressures in the radial and slit regions and another upstream of the slit (equivalent to P_1 in Figure 5.11) [49].

A. Exit Pressure Method

Consider the flow of a viscous polymer through a long slit die, as shown in Figure 5.12. As the polymer emerges into the atmosphere, it swells due to the presence of velocity rearrangements that are complete at section BB′. If we assume that fluid inertia can be neglected and that the velocity profile in the die is fully developed all the way to the exit, located at $x = L$, a momentum balance between sections AA′ and BB′ reveals that [50]

$$N_1(\dot{\gamma}_w) = P_e\left(1 + \frac{dlnP_e}{dln\tau_w}\right) \tag{5.7}$$

In Eq. (5.7), $\dot{\gamma}_w$ is the shear rate at the die wall, given by an equation similar to Eq. (2.2), τ_w is the shear stress at the die wall, and P_e is the exit pressure. The exit pressure is a nonzero value obtained by extrapolating, from the fully developed region,

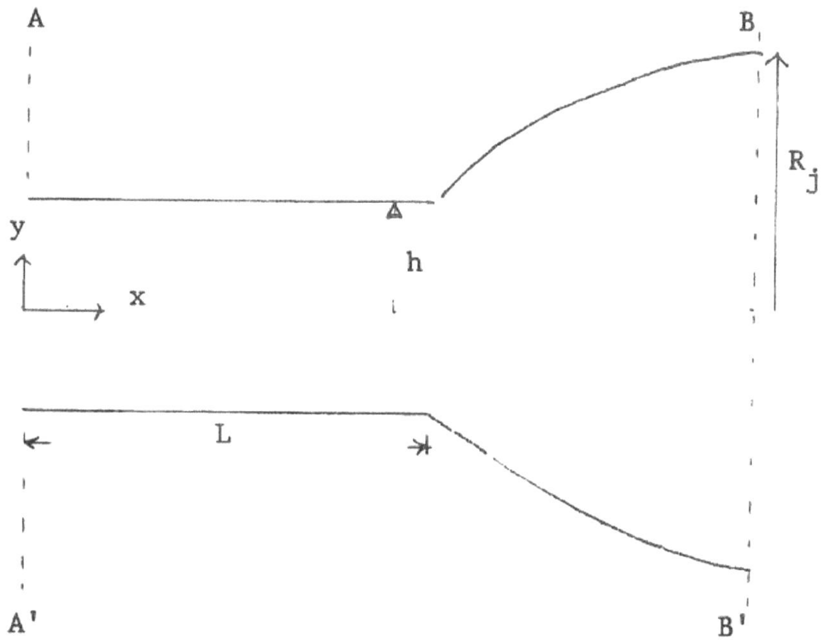

FIGURE 5.12 A polymeric fluid emerges into the atmosphere from a long slit die. A momentum balance is done between sections AA′ and BB′. (From Ref. [2].)

actual measurements of the wall normal stress (or pressure) made along the die wall to $x = L$; flush-mounted pressure transducers are used, and the pressure profile must be linear in x, and the extrapolation assumes that the flow is truly viscometric all the way to the die exit. This method of measuring N_1 has been extensively explored by Han [51 and references therein], who had helped commercialize the Seiscor/Han viscometer based on this principle. Typical data on a polystyrene melt are displayed in Figure 5.13, which also contains cone-and-plate data at low shear rates. It is seen that the exit pressure data points lie on a straight line obtained by extrapolating the rheometer results to higher shear rates. Furthermore, this straight-line plot is independent of the temperature of measurement and suggests that Eq. (5.3) remains valid to very high shear rates; similar conclusions are reached by examining data on other polymers [51].

In closing this subsection, we mention that Han has noted that Eq. (5.7) does not yield correct results if measurements are made at wall shear stresses below a critical value of about 25 kPa. This technique has been critically analyzed by Boger and Denn [50], who have questioned the validity of the fully developed velocity profile assumption. This, together with the need to differentiate experimental data, has led them to conclude that results obtained using Eq. (5.7) may only represent an upper bound on the value of N_1. Padmanabhan and Bhattacharya used two low-density polyethylene (LDPE) samples and compared N_1 results obtained with the help of the exit pressure method with similar results using the hole pressure method, and

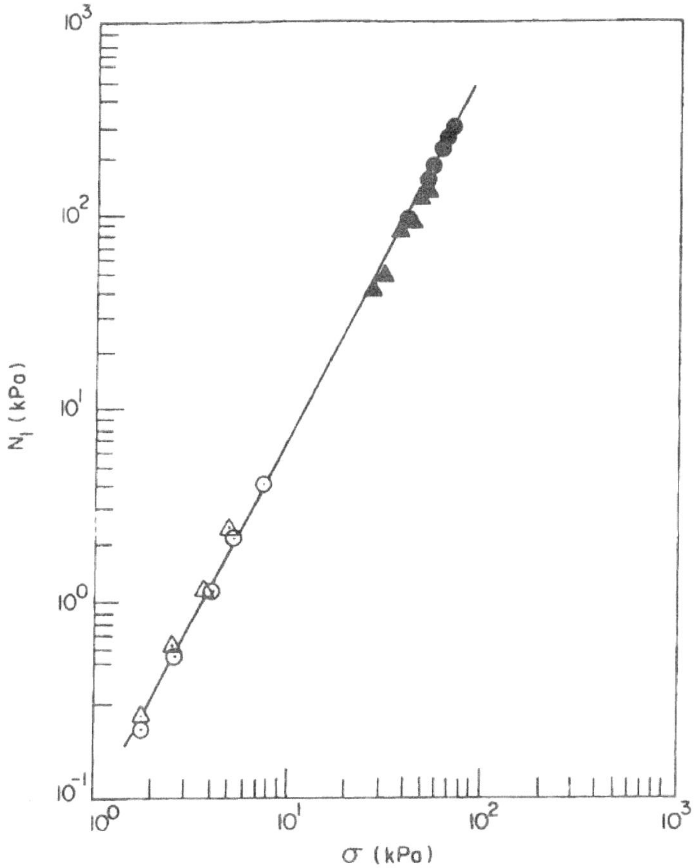

FIGURE 5.13 First normal stress difference versus shear stress for molten polystyrene at 200 and 220°C. Filled symbols are exit pressure data. (From Ref. [51].)

they recommend the use of the hole pressure method to avoid errors associated with extrapolation of data [18]; for corn meal extrusion, negative exit pressures and unreasonable results have been reported [52, 53].

B. Die Swell

This phenomenon, illustrated in Figure 5.4, is important in fabrication processes such as controlling the thickness of extruded sheets and in the making of bottles by extrusion blow molding. Shown in Figure 5.14 is the general behavior of the ratio of the jet diameter to the tube diameter as a function of the shear rate for polymer melts. Newtonian liquids and polymers at very low shear rates might be expected to have a die swell of 1.0, but actually they have a die swell of 1.12; this number can be computed by numerically solving the Navier–Stokes equation. Additional die swell first becomes noticeable at rates of shear at which the viscosity of the fluid

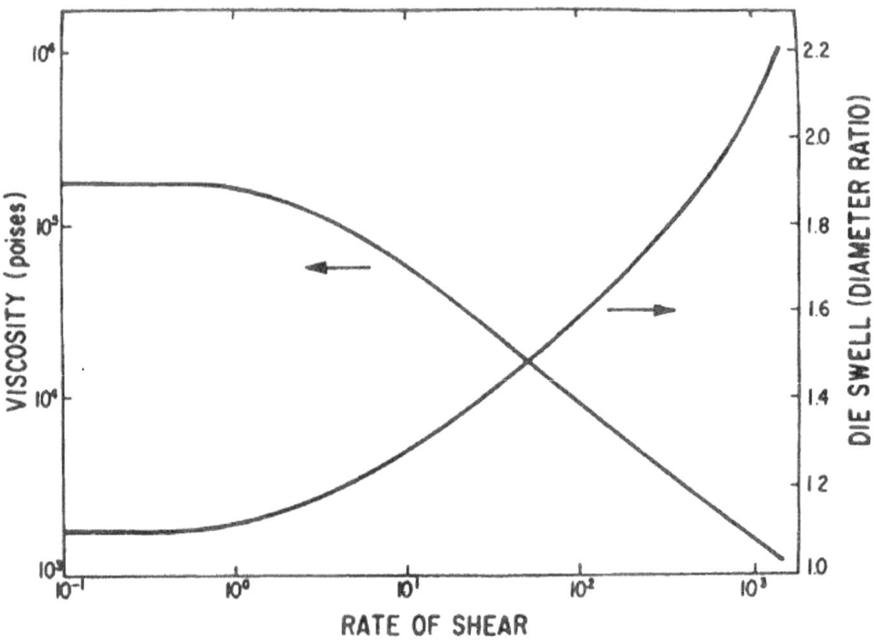

FIGURE 5.14 General behavior of die swell and viscosity as a function of the rate of shear in a capillary or similar duct. (From Ref. [12].)

starts to become non-Newtonian. The die swell increases with the rate of shear in a manner similar to the first normal stress difference, and this suggests that we might be able to determine N_1 from a measurement of die swell. However, die swell is not found to be a unique function of shear rate. For a given rate of shear, the die swell decreases as the length-to-diameter ratio of the capillary increases and ultimately attains an equilibrium value. This is due to the fact that extensional flow at the capillary entrance imparts greater molecular orientation compared to shear flow within the capillary. Thus, in a long capillary, some of the molecular orientation imparted to the polymer in the entrance region can relax out in the capillary itself. Also, at a constant shear rate, die swell (measured under isothermal conditions) tends to decrease somewhat as the temperature is raised because of the faster relaxation of molecular orientation at higher temperatures. But in analogy with the behavior of N_1, equilibrium die swell for a given polymer sample appears to be a unique function of the shear stress that is independent of the temperature of measurement [54]. Additionally, the variation of die swell with polymer structural variables such as molecular weight, molecular weight distribution, and long-chain branching seems to follow the same trends as in the case of the first normal stress difference [29, 30].

Various expressions have been proposed for relating die swell to the first normal stress difference, and several of these have been evaluated [30]. The most successful one appears to be based on a theory due to Tanner [55]. By assuming that the polymer leaving the tube is an elastic material that obeys Hooke's law, Tanner related die swell to the stresses acting on the polymer within the capillary. The final result is as follows:

$$\frac{D_j}{D} = 0.12 + \left(1 + \frac{S_R^{\,2}}{2}\right)^{1/6} \tag{5.8}$$

in which D_j is the jet diameter, D is the capillary diameter, S_R is recoverable shear, and the constant 0.12 is introduced to obtain agreement with Newtonian fluid behavior at low shear rates. If die swell is measured as a function of the wall shear stress, the first normal stress difference can be estimated with the help of Eq. (5.8). However, as noted by Middleman [54], due to the form of Eq. (5.8), small errors in measuring die swell can lead to large errors in the calculated value of the recoverable shear. In this context, it must be remembered that the die swell values required in Eq. (5.8) are isothermal values. It is found that swelling increases when a hot polymer is extruded through a tube with a colder wall. Thus, if extrusion is done in room-temperature air, it is necessary to anneal the extrudates to allow a complete recovery before measuring the jet diameter.

While many data sets agree with the predictions of Eq. (5.8), it has to be realized that this equation is a rather approximate one. Better predictions of die swell can be made using more realistic constitutive equations and more sophisticated mathematics, but the results show a great sensitivity to the nature of the viscoelastic constitutive equation employed [56]. Thus, one needs an accurate constitutive equation to relate die swell to N_1. However, if we had such an equation, we could use it directly to determine N_1; it would not be necessary to measure die swell at all!

C. Torsional Balance

In the parallel-plate viscometer described in Section V of Chapter 2, we use two parallel coaxial disks, separated by a fixed distance h, to shear the fluid of interest, and we determine a combination of N_1, and N_2 by measuring the force tending to pull the disks apart. In the *torsional balance* rheometer [57], we apply a known downward force to the upper disk and measure the resulting value of h as a function of the speed of rotation of the lower disk. Thus, Eqs. (2.17)–(2.19) still apply, and data analysis is done in the usual manner. The major advantage here is that smaller gap values can be attained, down to about 0.0003 cm, giving shear rates of the order of 10^5 sec^{-1}. In the conventional setup, such small gaps cannot be set with sufficient precision to obtain the same level of accuracy. Small gaps also reduce potential viscous heating problems.

This instrument has been used to measure the normal stresses in dilute polymer solutions, such as lubricants, but the two normal stress differences cannot be determined separately using this technique.

D. Hole Pressure Measurements

Unless flush-mounted pressure transducers are employed for flow through a slit die, the fluid pressure measured by drilling a hole through the channel wall will be

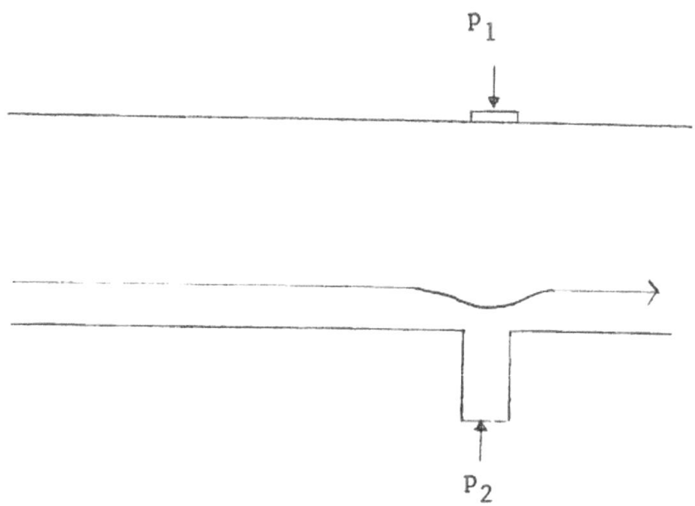

FIGURE 5.15 Influence of a pressure tap on streamline flow through a channel of rectangular cross-section. (From Ref. [2].)

lower than the true value. As shown in Figure 5.15, this hole pressure error arises due to the bending of streamlines toward the bottom of the hole. The reduction in pressure, $p_1 - p_2$, can be determined with the help of a flush-mounted transducer located at the same axial position but on the opposite wall of the slit. This pressure difference is found to be a monotonically increasing function of the first normal stress difference within the flowing liquid [58]. Indeed, it can be theoretically calculated if assumptions are made about the constitutive behavior of the fluid. In particular, for the steady, isothermal, creeping flow of a second-order fluid flowing over a transverse slot, the hole pressure error equals one-fourth of N_1 [59]. To analyze the situation for a more general viscoelastic fluid, several simplifying assumptions have to be made. According to Higashitani and Pritchard, the final results are [58, 60] as follows:

$$N_1 = 2n(p_1 - p_2) \tag{5.9}$$

where

$$n = \frac{d\log(p_1 - p_2)}{d\log\tau_w} \tag{5.10}$$

and, as expected, $n = 2$ for a second-order fluid.

Using the setup shown in Figure 5.11 to measure the hole pressure, Padmanabhan and Bhattacharya [18] applied this theory to report N_1 of commercial LDPE samples. The structure and properties of extruded starchy snacks, and hence their texture, are also determined by the elasticity of the material in its molten state [61].

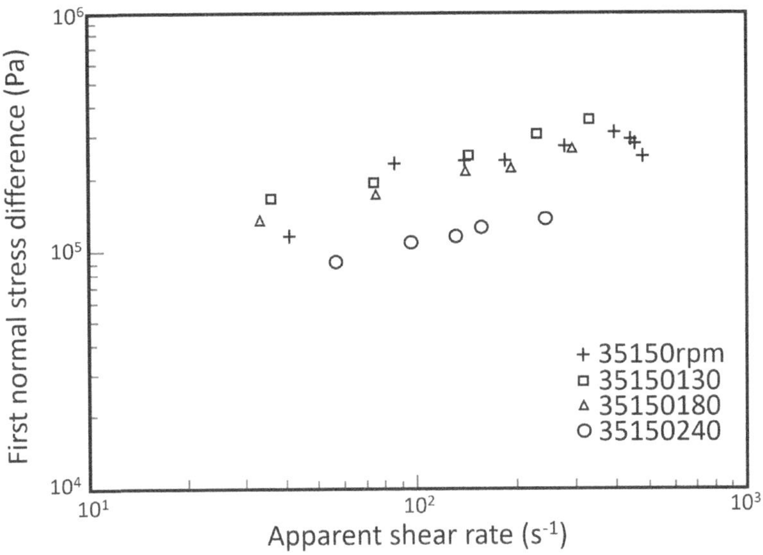

FIGURE 5.16 N_1 as a function of shear rate for corn meal at 35% moisture and 150°C from hole pressure measurements. (From Ref. [61].)

Bhattacharya and Padmanabhan used hole pressure measurements to determine N_1 of corn meal during extrusion cooking [53] using the same scheme as Figure 5.11 [18]. Note that the primary component of corn meal is starch, and the situation is similar to the extrusion of thermoplastic starch that was discussed in Chapter 3. Their results for N_1 as a function of shear rate are displayed in Figure 5.16; the corn meal contains 35% moisture, and the temperature of measurement is 150°C. Here, two different procedures were used: (1) in the points marked with a cross, the shear rate was changed by changing the extruder screw speed, and (2) for the other data sets, the screw speed was kept constant at 130, 180, or 240 rpm, and the shear rate was changed by changing the volumetric flow rate with the use of a bleed valve. The trends in each case are similar in that N_1 increases with increasing shear rate. However, certain other observations can be made. Procedure (1) subjects each point to a different processing history, and this is undesirable for N_1 measurements due to material degradation. Procedure (2) gives more consistent results, but data at the highest rpm (open diamonds) lie below the data from the other two screw speeds suggesting the effect of greater macromolecular degradation. It is noteworthy that the N_1 results for the corn meal are of the same order of magnitude as that of LDPE [18], both data having been measured using the same experimental setup.

The validity of the various assumptions in the Higashitani–Pritchard analysis and the modifications of this analysis have been discussed by Lodge [58]. A commercial instrument, employing hole pressure measurements and known as the Lodge Stressmeter, was built by the Bannatek Company in Madison, WI. It has been used to make measurements on multigrade motor oils at engine operating conditions—up to 10^6 sec^{-1} in shear rate and 150°C in temperature.

E. Jet Thrust Technique

In the momentum balance for the exit pressure method, Figure 5.12, fluid inertia could be neglected due to the very viscous nature of the liquids involved. For low-viscosity polymer solutions, especially at high Reynolds numbers, we can actually neglect the exit pressure but not the inertia of the fluid. If we also assume, as before, that the flow is fully developed all the way to the tube exit, the momentum balance now leads to (for a circular-cross-section tube of radius R) [50]

$$N_1(\dot{\gamma}_w)+0.5N_2(\dot{\gamma}_\omega)=\frac{1}{\pi R^2}\left(T_L+\frac{\tau_w}{2}\frac{\partial T_L}{\partial_{\tau_w}}\right) \tag{5.11}$$

in which T_L is the reduced thrust, defined as

$$T_L=2\pi\int_0^R \rho v^2(r,0)\,r\,dr-T \tag{5.12}$$

and T is the *jet thrust* on the tube. As long as all of the fluid enters the tube perpendicular to the jet, the magnitude of the thrust is $\rho\pi R_J^2\, v_J^2$; here, R_J is the jet radius and v_J its velocity. In practical terms, the jet thrust can be measured from the deflection of cantilever springs that carry the capillary tube. Thus, if jet thrust data are measured as a function of the wall shear stress, a combination of the two normal stress differences can be determined.

The history of the jet thrust method goes back to the very early 1960s. It has been a very popular technique due to its potential for providing high shear rate normal stress data on polymer solutions. Unfortunately, however, it now appears that the assumption of fully developed flow made in the derivation of Eq. (5.11) is in error. This makes the first normal stress difference calculated from jet thrust measurements higher than the true value by a substantial amount [62]. Since there is no theoretical or experimental way to estimate the extent of velocity rearrangement at the tube exit [50], the future utility of this method is in doubt.

IV. MEASUREMENT OF THE SECOND NORMAL STRESS DIFFERENCE

Compared to the volume of data available on the first normal stress difference, the amount of quantitative information about the second normal stress difference is quite meager. This is due primarily to two reasons. First, at a given shear rate, the size of N_2 is about an order of magnitude less than the size of N_1. Consequently, when a cone-and-plate viscometer is employed, either by itself or in combination with a parallel-plate instrument, the use of Eqs. (2.14)–(2.19) necessitates the taking of small differences between two large numbers, leading to very significant errors; one is often forced to conclude that N_2 is zero to within experimental error [63]; this result is also predicted by some theories in the dilute limit [1]. Second, it was thought that N_2 did not play much of a role in commercial polymer processing, and this fact resulted in minimal industrial interest in this material function [64]. Later, it was recognized that the accurate determination of the second normal stress difference may be important for both theoretical and practical reasons. On the theoretical side, such a measurement can help us in testing the quantitative validity of constitutive equations proposed for nonlinear viscoelasticity, while, on the practical side, N_2 may actually control important rheological phenomena. It has been suggested by Tanner and Keentok [65], and verified by Lee et al. [66], that edge fracture in a cone-and-plate viscometer occurs when N_2 exceeds a critical value. Also, Shaqfeh and coworkers have demonstrated that the second normal stress difference can strongly stabilize Taylor–Couette flow [67]. In addition, Levitt et al. have concluded that N_2 plays a significant role in morphology development during the preparation of polymer blends [68]. It is now well established that in flows through non-circular channels, secondary flows arise that result in fluid mixing [69]; this promotes both heat and mass transfer. Other flows where N_2 is important have been discussed by Maklad and Poole in their recent review on the subject [1].

A. Use of Standard Viscometers

According to Eq. (2.16), a graph of the total stress exerted by the fluid on the plate surface in a cone-and-plate viscometer when plotted against the logarithm of the radial position should be a straight line of slope ($N_1 + 2N_2$). This slope, together with an independent measurement of N_1, can, in principle, yield N_2 (one may also use Eq. (2.19) to obtain ($N_1 - N_2$) at the edge of a parallel-plate device). As mentioned earlier, this is not an easy measurement, and one may also need to make an inertia correction to the measured axial thrust value. Christiansen and coworkers used the procedure outlined here but employed a set of four extremely sensitive, custom-built, flush-mounted miniature pressure transducers in their cone-and-plate apparatus [33]. They also checked the internal consistency of all their data. Their second normal stress difference results on concentrated polystyrene solutions as a function of the shear stress are shown in Figure 5.17, and these resemble the functional behavior of N_1 versus shear stress seen earlier. In particular, the straight lines in Figure 5.17a have a slope of 2, and the average value of the ratio of $-N_2$ to N_1 is 0.25 (see Figure 5.17b). In addition, they found that for the monodisperse samples, N_2 increases with molecular weight to the 6.8–7.8 power; for polydisperse samples, N_2 increases with the width of the molecular weight distribution. It appears that long-chain branching also enhances the value of N_2 [70]. Lee and White [22] have published data on a number of molten polymers, and, while these data are relatively less accurate, they too show that N_2 is negative and considerably smaller than N_1. In addition, Meissner et al. [71] measured the ratio $-N_2/N_1$ for an LDPE at low shear rates using a cone-and-plate viscometer. They used a novel design in which the plate was partitioned into an inner disk connected to the normal force transducer while the outer ring was fixed to the frame of the instrument; this avoids issues of edge fracture. They found that the ratio of the two normal stress differences was 0.24. Other cone-and-plate geometries have also been described in the literature, and the reader is referred to the chapter by Powell for details [72].

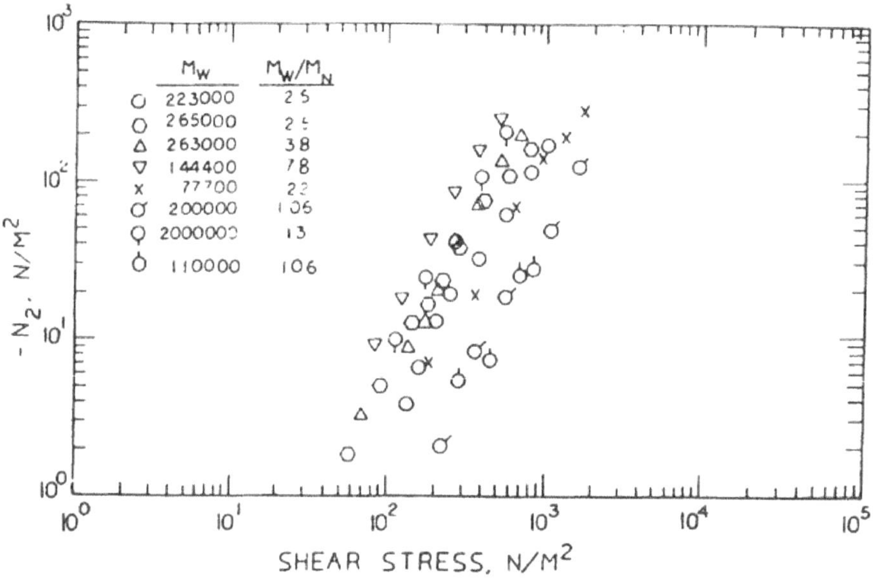

(a)

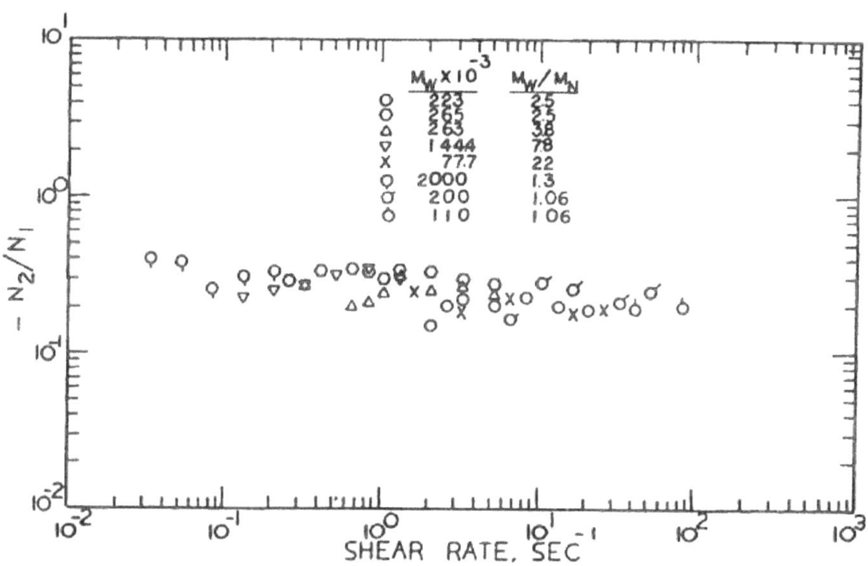

(b)

FIGURE 5.17 (a) The second normal stress difference versus shear stress for mono- and polydisperse polystyrene solutions in butylbenzene. (b) Normal stress difference ratio for these solutions. The concentration is 450 g/L except 250 g/L for the 2 million Mw polymer and 600 g/L for the 77,700 Mw polymer. (From Ref. [33].)

In closing this subsection, we mention that the principle of time–temperature superposition can be applied to second normal stress difference data to give a master curve using temperature shift factors obtained from the Williams, Landel, and Ferry (WLF) equation. This allows us to compare data obtained under different conditions from two different instruments or to extend the shear rate range of measurement at a given temperature [73].

B. Rod Climbing

As shown in Figure 5.5, polymeric liquids climb up a rotating shaft, and this effect has been put to good use by Beavers and Joseph [74]. These authors analyzed the Weissenberg effect and solved the equations of motion using a perturbation method to relate the height of climb h at an angular velocity ω to a climbing constant β defined as $0.5\psi_1^0 + \psi_2^0$; here ψ_1^0 and ψ_2^0 are, respectively, the zero-shear-rate values of the first and second normal stress coefficients. The final result is as follows:

$$h = h_s + \frac{4r\beta\omega^2}{2\sqrt{\rho g}\left(4\sqrt{\gamma} + r\sqrt{\rho g}\right)} \tag{5.13}$$

where h_s is the value of h in the absence of rotation, r is the rod radius, ρ is the liquid density, γ is the liquid surface tension, and g is the acceleration due to gravity. Magda et al. [75] have used this technique to determine the second normal stress difference of an organic Boger liquid [45] and found it to be negative and about 30 times smaller than the first normal stress difference. Although this is a simple method, it is limited to low shear rates, due to the assumptions made in the derivation of Eq. (5.13).

C. The Tilted Trough

If we consider gravity-driven flow down a semicircular cross-section inclined channel, we find that the fluid surface is flat if the fluid is Newtonian but is curved if the fluid is viscoelastic; a negative second normal stress difference results in a convex shape (bulging upwards), while a positive N_2 implies a concave surface (deflected downwards). Kuo and Tanner [76] have published an analysis that allows for the determination of N_2 as a function of the wall shear stress from a measurement of the surface profile. The theory, however, is somewhat involved, and the reader is referred to the original paper for details.

V. CONCLUDING REMARKS

In the past, accurate measurement of the normal stress differences of polymeric fluids in shear has been a fundamental problem in polymer rheology. As the techniques presented in this chapter reveal, much progress has been made in this regard, and a great deal is now known about the relationship between the normal stress differences and variables such as shear rate, temperature, molecular weight, molecular weight distribution, and long-chain branching. The experimental determination of N_1 is now considered a routine measurement useful for fluid characterization and the development of rheological models that might be utilized for process modeling and simulation. In this context, attempts have also been made to reverse the process and to use specific constitutive equations to predict the first normal stress difference function from the measured flow curve [77, 78]. Although some success has been achieved, the procedure cannot work in all cases. The Boger fluid, for example, has a constant viscosity, and it can be tailored to give different levels of elasticity [43]. Recall also the celebrated case of three polyethylenes that had the same viscosities and molecular weight distributions but behaved differently when used to make blown film [79]; the rheological differences among the polymers showed up in the first normal stress differences and in the extensional viscosities.

In the case of foods, despite the availability of commercial rheometers, very few attempts have been made to measure normal stress differences at low shear rates. At higher shear rates, present during processes such as extrusion, only few measurements of normal stress differences have been reported due to experimental difficulties. This is not to imply that these properties are not important.

In the future, we can expect more compact and user-friendly equipment with better software, an increased range of measurement, and greater accuracy and sensitivity. In addition, improved transducers should make it possible to measure the second normal stress difference with greater ease and to make reliable transient normal stress measurements, whether during step strain experiments or during startup and shutdown of shearing at a constant shear rate. Progress has also been made in the use of optical techniques for stress measurements. Fluid flow orients polymer molecules, and this makes the optical properties anisotropic. By relating the stress tensor to the refractive index tensor through the stress optic law, we can measure all the material functions with this noninvasive technique [80, 81]. Indeed, commercial viscometers are beginning to incorporate this feature, and researchers are increasingly using birefringence measurements along with mechanical stress measurements [82, 83].

REFERENCES

1. O. Maklad and R. J. Poole, A review of the second normal-stress difference; its importance in various flows, measurement techniques, results for various complex fluids and theoretical predictions. J. Non-Newt. Fluid Mech. 292:104522 (2021).
2. R. K. Gupta. Polymer and Composite Rheology. 2nd ed., Marcel Dekker, New York, 2000.
3. K. Walters. Rheometry. Chapman and Hall, London, 1975.
4. K. Oda, J. L. White, and E. S. Clark. Correlation of normal stresses in polystyrene melts. Polym. Eng. Sci. 18:25–28 (1978).
5. R. F. Ginn and A. B. Metzner. Measurement of stresses developed in steady laminar shearing flows of viscoelastic media. Trans. Soc. Rheol. 13:429–453 (1969).
6. R. J. Gordon and A. E. Everage Jr., Bead-spring model of dilute polymer solutions: Continuum modifications and an explicit constitutive equation. J. Appi. Polym. Sci. 15:1903–1909 (1971).
7. N. Phan-Thien. A nonlinear network viscoelastic model. J. Rheol. 22:259–283 (1978).
8. J. Meissner. Modifications of the Weissenberg rheogoniometer for measurement of transient rheological properties of molten polyethylene under shear. Comparison with tensile data. J. Appi. Polym. Sci. 16:2877–2899 (1972).

9. R. B. Bird, R. C. Armstrong, and O. Hassager. Dynamics of Polymeric Liquids. 2nd ed. Vol. 1. Wiley, New York, 1987.
10. M. H. Wagner. Analysis of time-dependent non-linear stress-growth data for shear and elongational flow of a low-density branched polyethylene melt. Rheol. Acta 15:136–142 (1976).
11. E. V. Menezes and W. W. Graessley. Nonlinear rheological behavior of polymer systems for several shear-flow histories. J. Polym. Sci.: Polym. Phys. Ed. 20:1817–1833 (1982).
12. L. E. Nielsen. Polymer Rheology. Marcel Dekker, New York, 1977.
13. A. S. Lodge. Elastic Liquids. Academic Press, New York, 1964.
14. B. D. Coleman, H. Markovitz, and W. Noll. Viscometric Flows of Non-Newtonian Fluids. Springer-Verlag, New York, 1966.
15. E. B. Christiansen and W. R. Leppard. Steady-state and oscillatory flow properties of polymer solutions. Trans. Soc. Rheol. 18:65–86 (1974).
16. Z. Tadmor and R. B. Bird. Rheological analysis of stabilizing forces in wire-coating dies. Polym. Eng. Sci. 14:124–136 (1974).
17. B. Maxwell and A. J. Scalora. The elastic melt extruder—Works without screw. Modern Plastics 37:107, October 1959.
18. M. Padmanabhan and M. Bhattacharya. In-line measurements of rheological properties of polymer melts. Rheol. Acta 33:71–87 (1994).
19. D. B. Genovese and M. A. Rao. Apparent viscosity and first normal stress difference of starch dispersions: Role of continuous and dispersed phases, and prediction with the Goddard-Miller model. Appl. Rheol. 13:183–190 (2003).
20. N. E. Hudson and J. Ferguson. The shear flow properties of M1. J. Non-Newt. Fluid Mech. 35:159–168 (1990).
21. D. A. Nguyen and T. Sridhar. Preparation and some properties of M1 and its constituents. J. Non-Newt. Fluid Mech. 35:93–104 (1990).
22. B. C. Lee and J. L. White. An experimental study of rheological properties of polymer melts in laminar shear flow and of interface deformation and its mechanisms in two-phase stratified flow. Trans. Soc. Rheol. 18:467–492 (1974).
23. W. Minoshima, J. L. White, and J. E. Spruiell. Experimental investigation of the influence of molecular weight distribution on the rheological properties of polypropylene melts, Polym. Eng. Sci. 20:1166–1176 (1980).
24. J. Greener and G. H. Pearson. Orientation residual stresses and birefringence in injection molding. J. Rheol. 27:115–134 (1983).
25. J. Vlachopoulos and N. Polychronopoulos. Basic concepts in polymer melt rheology and their importance in processing. In: M. Kontopoulou, ed. Applied Polymer Rheology: Polymeric Fluids with Industrial Applications. Wiley, Hoboken, NJ, 2012, pp. 1–27.
26. C. J. S. Petrie and M. M. Denn. Instabilities in polymer processing. AIChE J. 22:209–236 (1976).
27. A. B. Metzner, J. L. White, and M. M. Denn. Behavior of viscoelastic materials in short-time processes. Chem. Eng. Prog. 62:81–92 (December 1966).
28. H. M. Laun. Prediction of elastic strains of polymer melts in shear and elongation. J. Rheol. 30:459–501 (1986).
29. R. Racin and D. C. Bogue. Molecular weight effects in die swell and in shear rheology. J. Rheol. 23:263–280 (1979).
30. C. D. Han. Rheology in Polymer Processing. Academic Press, New York, 1976.
31. R. A. Mendelson, W. A. Bowles, and F. L. Finger. Effect of molecular structure on polyethylene melt rheology. I. Low-shear behavior. J. Polym. Sci.: Part A-2 8:105–126 (1970).
32. J. L. White and H. Yamane. A collaborative study of the rheological properties and unstable melt spinning characteristics of linear and branched polyethylene terephthalates. Pure Appi. Chem. 57:1441–1452 (1985).

33. H. W. Gao, S. Ramachandran, and E. B. Christiansen. Dependence of the steady-state and transient viscosity and first and second normal stress difference functions on molecular weight for linear mono- and polydisperse polystyrene solutions. J. Rheol. 25:213–235 (1981).
34. R. I. Tanner. A correlation of normal stress data for polyisobutylene solutions. Trans. Soc. Rheol. 17:365–373 (1973).
35. M. Oritz, D. De Kee, and P. J. Carreau. Rheology of concentrated poly(ethylene oxide) solutions, J. Rheol. 38:519–539 (1994).
36. A. V. Shenoy and D. R. Saini. An approach to the estimation of polymer melt elasticity. Rheol. Acta 23:608–616 (1984).
37. G. Astarita and G. Marrucci. Principles of Non-Newtonian Fluid Mechanics. McGraw Hill, London, 1974.
38. R. G. Larson. Constitutive Equations for Polymer Melts and Solutions. Butterworths, Boston, 1988.
39. R. B. Bird, H. R. Warner Jr., and D. C. Evans. Kinetic theory and rheology of dumbbell suspensions with Brownian motion. Adv. Polym. Sci. 8:1–90 (1971).
40. A. Kumar and R. K. Gupta. Fundamentals of Polymer Engineering. 3rd ed., CRC Press, Boca Raton, FL, 2019.
41. M. J. Crochet, A. R. Davies, and K. Walters. Numerical Simulation of Non-Newtonian Flow. Elsevier, Amsterdam, 1984.
42. K. C. Tam and C. Tiu. A low viscosity, highly elastic ideal fluid. J. Non-Newt. Fluid Mech. 31:163–177 (1989).
43. K. C. Tam, T. Moussa, and C. Tiu. Ideal elastic fluids of different viscosity and elasticity levels. Rheol. Acta 28:112–120 (1989).
44. D. V. Boger. A highly elastic constant-viscosity fluid. J. Non-Newt. Fluid Mech. 3: 87–91 (1977/78).
45. G. Prilutski, R. K. Gupta, T. Sridhar, and M. E. Ryan. Model viscoelastic liquids. J. Non-Newt. Fluid Mech. 12:233–241 (1983).
46. R. B. Bird, C. F. Curtiss, R. C. Armstrong, and O. Hassager. Dynamics of Polymeric Liquids. 2nd ed. Vol. 2. Wiley, New York, 1987.
47. M. Doi and S. F. Edwards. The Theory of Polymer Dynamics. Oxford University Press, Oxford, 1986.
48. M. Doi. Introduction to Polymer Physics. Oxford University Press, Oxford, 1996.
49. M. K. Esfahani, C. K. Georgantopoulos, I. F. C. Naue, J. Sunder, and M. Wilhelm. A new slit-radial die for simultaneously measuring steady state shear viscosity and first normal stress difference of viscoelastic liquids via capillary rheometry. J. Appl. Polym. Sci. 139:e52094 (2022).
50. D. V. Boger and M. M. Denn. Capillary and slit methods of normal stress measurements. J. Non-Newt. Fluid Mech. 6:163–185 (1980).
51. C. D. Han. Slit rheometry. In: A. A. Colyer and D. W. Clegg, eds. Rheological Measurement, 2nd ed. Chapman & Hall, London, 1998, pp. 190–209.
52. M. Padmanabhan and M. Bhattacharya. Flow behavior and exit pressures of corn meal under high-shear-high-temperature extrusion conditions using a slit die. J. Rheol. 35:315–343 (1991).
53. M. Bhattacharya and M. Padmanabhan. Evaluation of the hole pressure method to measure the first normal stress difference of corn meal dough during extrusion cooking. J. Text. Studies 25:241–265 (1994).
54. S. Middleman. Fundamentals of Polymer Processing. McGraw Hill, New York, 1977.
55. R. I. Tanner. A theory of die-swell. J. Polym. Sci.: Part A-2 8:2067–2078 (1970).
56. R. I. Tanner. Recoverable elastic strain and swelling ratio. In: A. A. Collyer and D. W. Clegg, eds. Rheological Measurement, 2nd ed., Chapman & Hall, London, 1998, pp. 492–515.

57. D. M. Binding and K. Walters. Elastico-viscous squeeze films, part 3. The torsional balance rheometer. J. Non-Newt. Fluid Mech. 1:277–286 (1976).
58. A. S. Lodge. Normal stress differences from hole pressure measurements. In: A. A. Collyer and D. W. Clegg, eds. Rheological Measurement, 2nd ed., Chapman & Hall, London, 1998, pp. 299–326.
59. R. L. Tanner and A. C. Pipkin. Intrinsic errors in pressure-hole measurements. Trans. Soc. Rheol. 13:471–484 (1969).
60. K. Higashitani and W. G. Pritchard. A kinematic calculation of intrinsic errors in pressure measurement made with holes. Trans. Soc. Rheol. 16:687–696 (1972).
61. M. Padmanabhan and M. Bhattacharya. Rheological measurement of fluid elasticity during extrusion cooking. Trends Food Sci. Tech. 2:149–151 (1991).
62. J. M. Davies, J. F. Hutton, and K. Walters. A critical re-appraisal of the jet-thrust technique for normal stresses, with particular reference to axial velocity and stress rearrangement at the exil plane. J. Non-Newt. Fluid Mech. 3:141–160 (1977/78).
63. K. Walters. The second-normal-stress difference project. IUPAC Macro-83. Bucharest, Romania. Plenary and invited lectures. Part 2, IUPAC Macromolecular Division, 1983, pp. 227–237.
64. K. Walters. Fundamental concepts. In: K. Walters, ed. Rheometry: Industrial Applications. Research Studies Press, Chichester, 1980, pp. 1–29.
65. R. I. Tanner and M. Keentok. Shear fracture in cone-plate rheometry. J. Rheol. 27:47–57 (1983).
66. C. S. Lee, B. C. Tripp, and J. J. Magda. Does N_1 or N_2 control the onset of edge fracture? Rheol. Acta 31:306–308 (1992).
67. E. S. G. Shaqfeh, S. J. Muller, and R. G. Larson. The effects of gap width and dilute solution properties on the viscoelastic Taylor-Couette instability. J. Fluid Mech. 235:285–317 (1992).
68. L. Levitt, C. Macosko, and S. D. Pearson. Influence of normal stress difference on polymer drop deformation. Polym. Eng. Sci. 36:1647–1655 (1996).
69. B. Debbaut, T. Avalosse, J. Dooley, and K. Hughes. On the development of secondary motions in straight channels induced by the second normal stress difference: Experiments and simulations. J. Non-Newt. Fluid Mech. 69:255–271 (1997).
70. C.-S. Lee, J. J. Magda, and K. L. DeVries. Measurements of the second normal stress difference for star polymers with highly entangled branches. Macromolecules 25:4744–4750 (1992).
71. J. Meissner, R. W. Garbella, and J. Hostettler. Measuring normal stress differences in polymer melt shear flow. J. Rheol. 33:843–864 (1989).
72. R. L. Powell. Rotational viscometry. In: A. A. Colyer and D. W. Clegg, eds. Rheological Measurement, 2nd ed., Chapman & Hall, London, 1998, pp. 260–298.
73. G. A. Alvarez, A. S. Lodge, and H.-J. Cantow. Measurement of the first and second normal stress differences: Correlation of four experiments on a polyisobutylene/decalin solution "DI". Rheol. Acta 24:368–376 (1985).
74. G. S. Beavers and D. D. Joseph. The rotating rod viscometer. J. Fluid Mech. 69:475–511 (1975).
75. J. J. Magda, J. Lou, S. G. Baek, and K. L. DeVries. Second normal stress difference of a Boger fluid. Polymer 32:2000–2009 (1991).
76. Y. Kuo and R. I. Tanner. On the use of open-channel flows to measure the second normal stress difference. Rheol. Acta 13:443–456 (1974).
77. S. I. Abdel-Khalik, O. Hassager, and R. B. Bird. Prediction of melt elasticity from viscosity data. Polym. Eng. Sci. 14:859–867 (1974).
78. M. H. Wagner. Predictions of primary normal stress difference from shear viscosity data using a single integral constitutive equation. Rheol. Acta 16:43–50 (1977).
79. J. Meissner. Basic parameters, melt rheology, processing and end-use properties of three similar low density polyethylene samples. Pure Appi. Chem. 42:553–612 (1975).

80. J. L. S. Wales. The Application of Flow Birefringence to Rheological Studies of Polymer Melts. Delft University Press, Rotterdam, 1976.
81. H. Janeschitz-Kriegl. Polymer Melt Rheology and Flow Birefringence. Springer-Verlag, Berlin, 1983.
82. E. F. Brown, W. R. Burghardt, H. Kahvand, and D. C. Venerus. Comparison of optical and mechanical measurements of second normal stress difference relaxation following step strain. Rheol. Acta 34:221–234 (1995).
83. T. Takahashi, M. Shirakashi, K. Miyamoto, and G. G. Fuller. Development of a double beam rheo-optical analyzer for full tensor measurement of optical anisotropy in complex fluid flow. Rheol. Acta 41:448–455 (2002).

6 Dynamic Mechanical Properties

I. INTRODUCTION

A dynamic mechanical experiment is one in which a polymer sample is subjected to a sinusoidal strain γ of infinitesimal amplitude γ_0 and fixed angular frequency ω; $\omega = 2\pi f$, where f is the frequency in hertz. Therefore,

$$\gamma = \gamma_0 \sin \omega t \tag{6.1}$$

Here, the stress response τ is found to be sinusoidal but, in general, is out of phase by an angle δ. Thus,

$$\tau = \tau_0 \sin(\omega t + \delta) \tag{6.2}$$

or

$$\tau = (\tau_0 \cos \delta) \sin \omega t + (\tau_0 \sin \delta) \cos \omega t \tag{6.3}$$

This is known as linear behavior, and it is observed regardless of the mode of deformation, whether shear, extension, or flexure. Linear behavior means that the ratio of the stress to the strain does not depend on the strain amplitude γ_0; the maximum strain amplitude for which Eq. (6.2) holds cannot be predicted *a priori* but has to be determined by experiment. On dividing the stress by the strain amplitude, we get the modulus G:

$$G = G'(\omega) \sin \omega t + G''(\omega) \cos \omega t \tag{6.4}$$

where $G' = (\tau_0 \cos \delta)/\gamma_0$ and $G'' = (\tau_0 \sin \delta)/\gamma_0$. The term G', called the *storage or elastic modulus*, is the same quantity defined by Eq. (1.7), and it is the in-phase component of the modulus and represents the energy stored and recovered per cycle [1]. Correspondingly, the term G'', called the *loss or viscous modulus*, was defined earlier by Eq. (1.8), and it is the out-of-phase component of the modulus and represents the energy dissipated as heat per cycle of deformation. The ratio of G'' to G' is *tan* δ and is an alternate measure of energy dissipation.

The basic essentials of most instruments used for measuring dynamic properties are schematically illustrated in Figure 6.1. The polymer sample is deformed in shear or in tension by some oscillating driver, which may be either mechanical or electromagnetic in nature. The amplitude of the sinusoidal deformation is measured by a strain transducer, which may be a linear variable-differential transformer (LVDT),

DOI: 10.1201/9781003013501-6

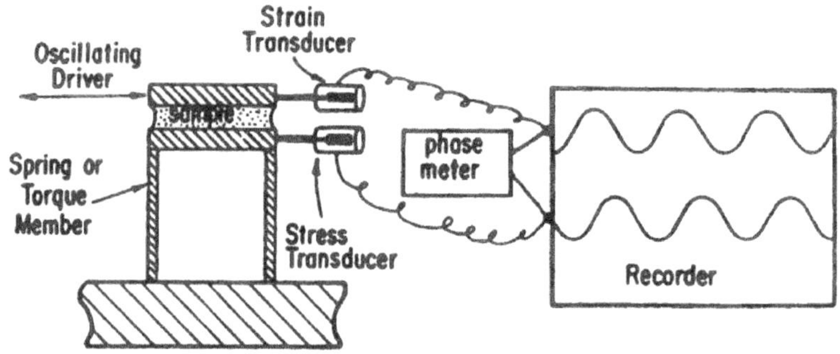

FIGURE 6.1 Schematic diagram of an oscillating rheometer for measuring the dynamic properties of fluids. (From Ref. [2].)

a variable-resistance gage, or some type of optical transducer. The force deforming the sample is measured by the small deformation of a relatively rigid spring or torsion bar to which is attached a stress transducer; this may be another LVDT. A few of the commercial instruments and others that are widely used are described in the Refs. [1–6]. A special type of *oscillating rheometer*, known as the *orthogonal rheometer*, consists of two parallel disks, the space between which is filled with polymer, and whose axes are not quite collinear (see Figure 2.4b). As one disk rotates relative to the other, an eccentric oscillatory motion is set up in the polymer [7].

Dynamic mechanical experiments are conducted routinely on both polymeric solids and polymeric liquids, and these are useful for material characterization and quality control purposes. The experiments may be conducted as a function of temperature at a fixed frequency or as a function of frequency under isothermal conditions. The former mode of operation is particularly well suited for work on solids, and Figure 6.2 shows representative data on an (amorphous) sample of polymethyl methacrylate subjected to torsional deformation at a frequency of 1 Hz. At low temperatures, the polymer is glassy, and it acts essentially as an elastic solid; this is confirmed by the fact that $G' \gg G''$. Under these conditions, the storage modulus equals the shear modulus, or stiffness; this value is typically on the order of 10^9 Pa. As the temperature approaches the glass transition temperature, approximately 125°C in the present case, the storage modulus falls rapidly by about three orders of magnitude while the loss modulus and tan δ increase rapidly and exhibit maxima. Thus, these data can be used to identify thermal transitions such as the glass transition temperature and secondary dispersions involving the motion of side chains of the polymer. Indeed, this technique has been used extensively to probe the morphology of semicrystalline polymers [3, 8, and references therein]. Note, though, that T_g measured using dynamic mechanical analysis is usually slightly larger than that measured with the help of a differential scanning calorimeter; typically, T_g increases with increasing frequency or a reduction in the time scale of measurement [3].

For molten polymers, it is more common to measure the storage and loss moduli as a function of frequency while keeping temperature unchanged. The characteristic response of a high-molecular-weight molten polymer to small-amplitude oscillatory

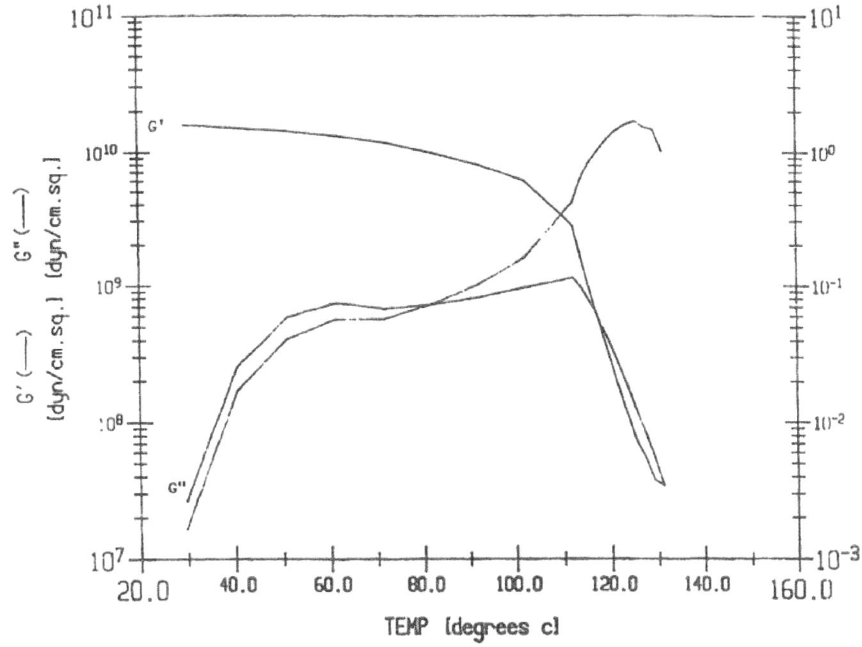

FIGURE 6.2 Storage modulus, loss modulus, and damping factor as a function of temperature for a polymethyl methacrylate sample subjected to torsional deformation at 1 Hz. (From Ref. [9].)

shear deformation over 12 decades of frequency is sketched in Figure 6.3a [10]. It is seen that at the lowest frequencies (corresponding to small values of the Deborah number), the response is liquidlike, with the loss modulus exceeding the storage modulus. However, in this region, known as the *terminal region*, the double logarithmic curve of G' has a slope of 2 while that of G'' has a slope of unity. Consequently, the two moduli soon become equal and display a plateau region. This is typical rubbery behavior and arises from the presence of chain entanglements; the magnitude of G' in the plateau region is known as the *plateau modulus* G_N° (Figure 6.3a). At still higher frequencies, there is a *transition region*, where the polymer feels leathery, that gives way to glassy behavior. Here, the storage modulus becomes constant while the loss modulus goes through a maximum. The melt acts like a glassy solid at very high frequencies (corresponding to large values of the Deborah number) because it is only the small parts of the molecule that can respond over such short time scales. As we shall see later, dynamic mechanical data can be used to predict other small-strain properties of the polymer, and this is one major reason why we carry out oscillatory experiments. The truth of this assertion can actually be seen by comparing storage modulus versus frequency data with the *stress relaxation modulus* as a function of time. In a *stress relaxation experiment*, a constant strain is imposed on the polymer, and the time-dependent stress is monitored; Figure 6.3b displays the result [10]. Clearly, Figures 6.3a and 6.3b are similar, and one can be obtained from the other by equating time to the reciprocal of the frequency. Note that the frequency at which the

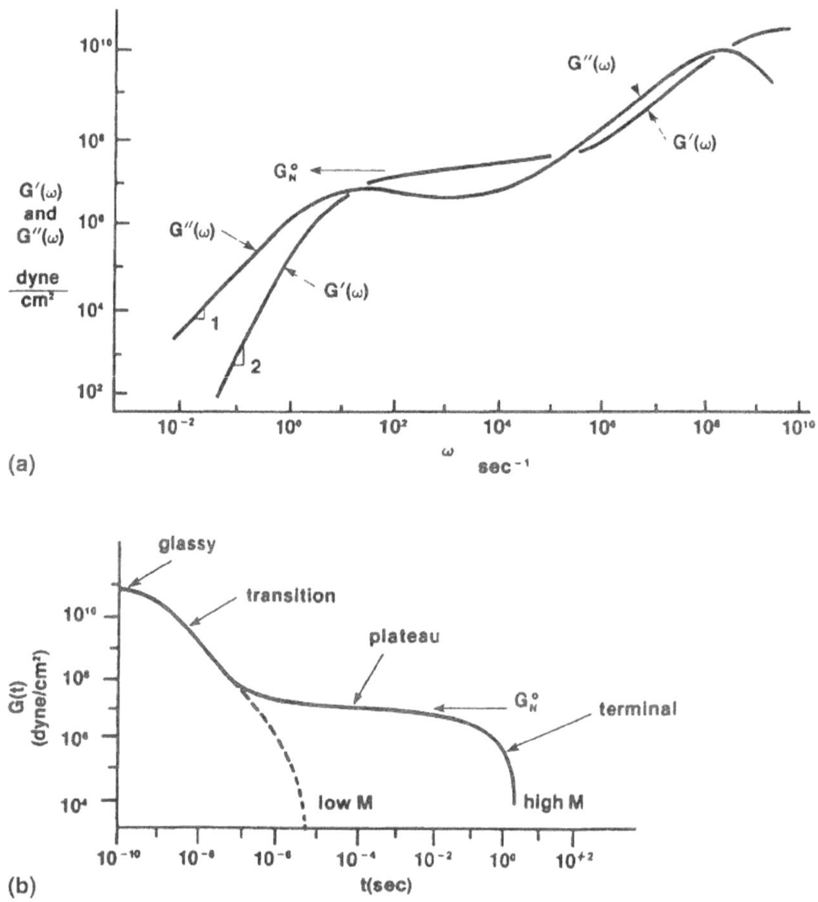

FIGURE 6.3 (a) Isothermal storage and loss moduli of a molten polymer as a function of frequency, (b) Stress relaxation behavior of the same molten polymer. (Reprinted with permission from Ref. [10]. Copyright 1984 American Chemical Society.)

G' and G'' curves first intersect each other is known as the *crossover frequency*, and the two moduli equal G_c. The crossover frequency is supposed to separate regions of viscous behavior from regions of elastic behavior, and it signifies the beginning of the elastic plateau region [11]. For polymer solutions, the crossover frequency increases as the concentration decreases. It has also been postulated in the literature that the polydispersity index of the polymer is approximately equal to $10^5/G_c$ when G_c is measured in Pa [12, 13]. This is especially true for polypropylene melts. Indeed, G_c is found to be a decreasing function of the polydispersity index, but when the polydispersity index is very large (very broad molecular weight distribution), G_c is found to correlate better with higher moments of the molecular weight distribution [14]. This conclusion is based on data for high-density polyethylene, whether made using Ziegler-Natta catalysts or metallocene catalysts.

II. DEPENDENCE OF POLYMER MELT DYNAMIC MECHANICAL PROPERTIES ON TEMPERATURE, MOLECULAR WEIGHT, AND POLYMER STRUCTURE

Using commercial instruments, it is generally not possible to obtain dynamic data over the wide frequency range illustrated in Figure 6.3a. Instead, we make measurements at different constant temperatures and then employ the principle of time–temperature superposition to construct a master curve at a chosen reference temperature; this procedure has been described in detail in Section III.B of Chapter 3. Shown in Figure 6.4 are isothermal, storage modulus data as a function of angular frequency for a 167,000-molecular-weight, narrow-distribution polystyrene melt at a variety of temperatures [15]. These data can be made to collapse into a single curve by means of a horizontal shift, and the result is displayed in Figure 6.5 at a reference temperature of 160°C. Figure 6.5 also shows similar master curves for other molecular weight fractions, ranging from 8900 (curve L9) to 581,000 (curve LI8) [15]. Corresponding data for the loss modulus are given in Figure 6.6 [15]. The values of the shift factors needed to obtain these master curves depend on temperature alone, and these can be determined using the WLF equation [Eqs. (3.9) and (3.19)]. Clearly, changes in temperature are equivalent to changes in frequency or time.

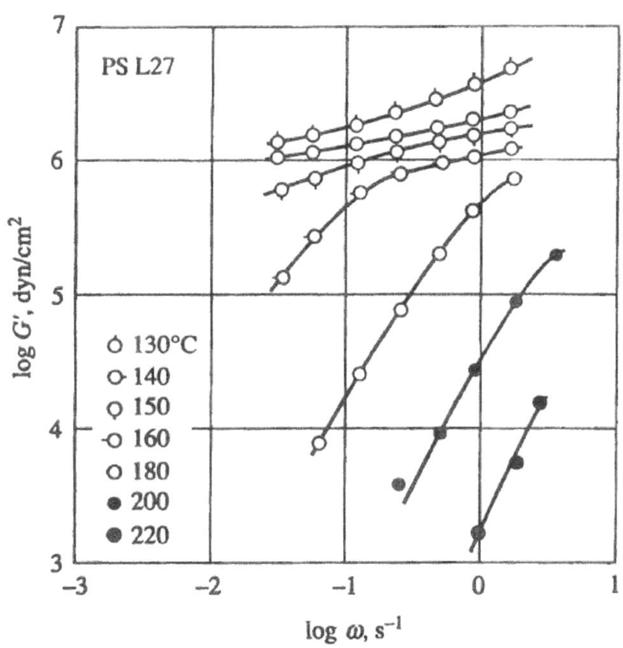

FIGURE 6.4 Frequency dependence of G' for narrow-distribution polystyrene L27, molecular weight 167,000, at various temperatures. (Reprinted with permission from Ref. [15]. Copyright 1970 American Chemical Society.)

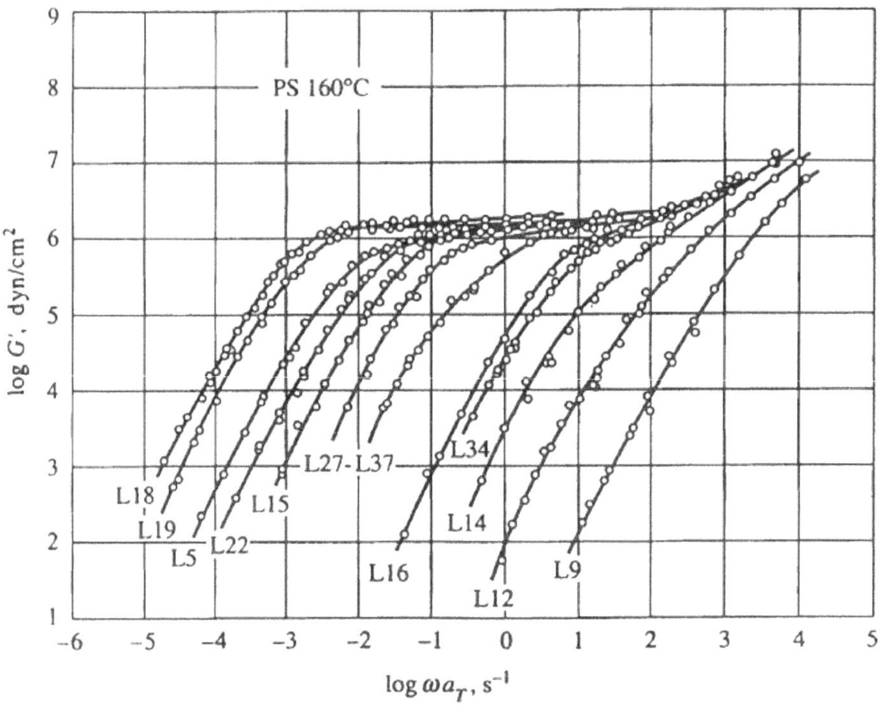

FIGURE 6.5 Master curves of G' for narrow-distribution polystyrenes having different molecular weights. The reference temperature is 160°C. (Reprinted with permission from Ref. [15]. Copyright 1970 American Chemical Society.)

Figures 6.5 and 6.6 reveal that both the elastic modulus and the loss modulus increase rapidly with frequency and with molecular weight up to a plateau value of about 10^5 Pa. Increasing the polymer molecular weight shifts the flow or terminal region to progressively smaller frequencies. Low-molecular-weight polymers that have few if any entanglements have little or no plateau region in the log G' or log G'' versus log ωa_T plots. High-molecular-weight polymers have plateau regions that cover several decades of frequency. While the G' curve exhibits a sharp transition from the *rubbery* to the terminal zone, the G'' curve goes through a maximum. These plateaus exist in a frequency range above which the entanglements do not have time to slip and relax the stress. These values of G' and G'' in the plateau regions are essentially independent of molecular weight, because the effective molecular weight between entanglements becomes constant. Entanglements that cannot relax out behave as crosslinks, and storage modulus data together with the kinetic theory of rubber elasticity can be used to estimate the molecular weight between entanglements [16].

The upturn in all the curves in Figures 6.5 and 6.6 at reduced frequencies greater than 10^3 radians/second is due to the beginning of the glass transition region. The various curves, except those corresponding to the lowest molecular weights, appear

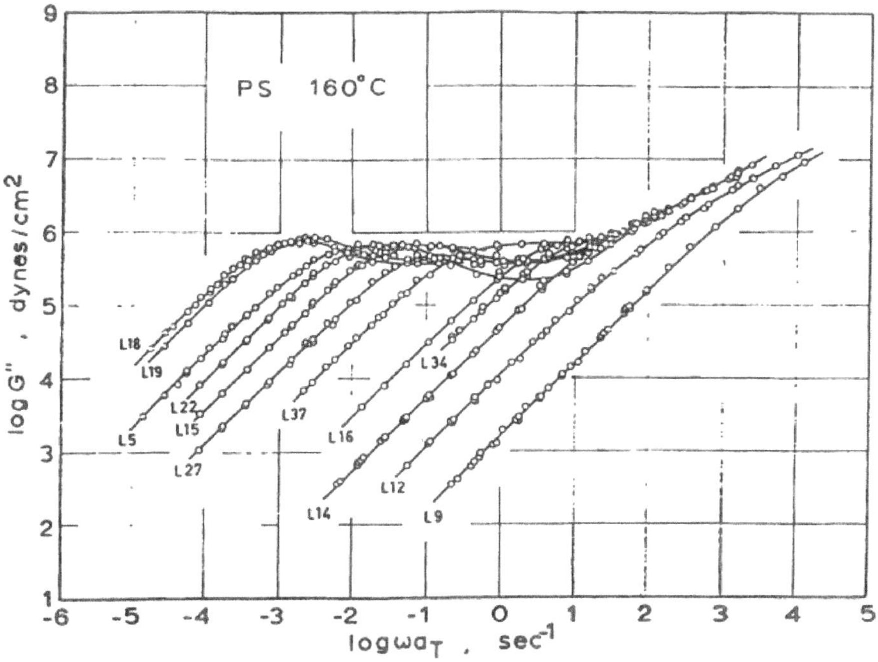

FIGURE 6.6 Master curves of G'' for narrow-distribution polystyrenes having different molecular weights. The reference temperature is 160°C. (Reprinted with permission from Ref. [15]. Copyright 1970 American Chemical Society.)

to come together here. The polymer melt no longer behaves as a fluid but begins to take on the characteristics of a glass at these high frequencies, and many kinds of molecular motions are no longer able to take place in time intervals less than 10^{-3} seconds.

On going from monodisperse to polydisperse polymers, we find that the shapes of the G' and G'' functions change considerably in the terminal and plateau regions but not in the transition region. In particular, as the width of the molecular weight distribution increases, the changeover from the rubbery to the terminal region in the storage modulus curve becomes much more gradual due to the presence of high-molecular-weight fractions, and even the plateau disappears [17]; this is seen in Figure 6.7 [18]. Simultaneously, G'' loses the peak that monodisperse polymers have in the frequency range in which G' tends to become constant [17]. The introduction of long-chain branching generally decreases the elasticity or G' of the melts [18, 19]; typical data are again shown in Figure 6.7, which compares the behavior of narrow-distribution, broad-distribution, and four-branch star polystyrenes [18]. Measurements on narrow molecular weight distribution metallocene polyethylenes, however, show that, as the long-chain branching density is increased, fluid elasticity increases due to entanglement formation. One manifestation of this is a progressive increase in the storage modulus, especially at low frequencies [20].

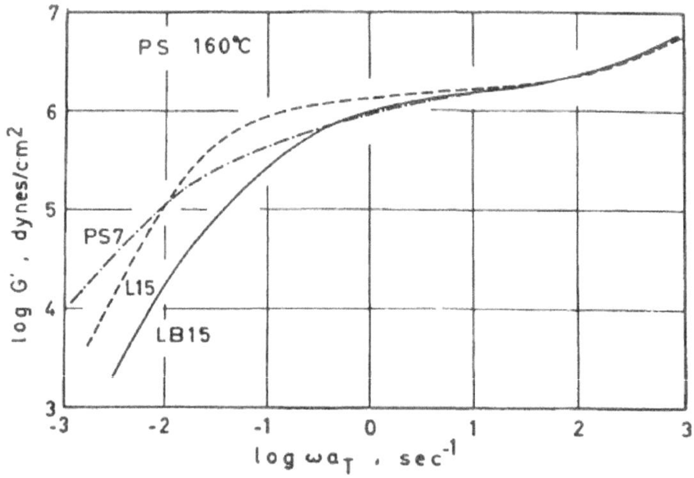

FIGURE 6.7 A comparison of G' curves for three different polystyrenes: L15 is monodisperse and linear, PS7 is broad distribution and linear, and LB15 is a four-branch star. The weight-average molecular weight in each case is similar. (Reprinted with permission from Ref. [18]. Copyright 1971 American Chemical Society.)

The fairly strong influence that polydispersity has on the dynamic mechanical properties of linear polymers has prompted efforts aimed at determining the molecular weight distribution from linear viscoelasticity data [21, 22]. It is well known that conventional methods of molecular weight distribution measurement require that the polymer sample be soluble in some solvent at ambient temperatures [23]. However, many industrially important polymers, especially fluoropolymers, are fairly insoluble in common solvents; this has been one major motivation for the development of rheological methods for molecular weight distribution determination [24]. To be successful in this endeavor, we need mixing rules that relate the composition and properties of the components in a polydisperse sample to the measured behavior of the mixture. Such mixing rules can be derived with the help of rheological constitutive equations, but the mathematics can be quite formidable. Consequently, the reader is referred to the technical literature for details [20, 22]. The method, though, appears to work, and commercial instruments now come equipped with appropriate software.

Just as data for the first normal stress difference in shear lie on a temperature-independent line when plotted against the shear stress on double logarithmic coordinates (see Figures 5.7 and 5.8), it is found that similar, temperature-independent correlations arise on plotting the logarithm of G'' versus the logarithm of G' [25]. This is true even for miscible polymer blends; absence of superposition on changing blend composition is an indication of rheological incompatibility [26]. For polyethylenes, the absence of superposition is related to the presence of long-chain branching, and this is one way to independently distinguish between the effects of polydispersity and long-chain branching [27].

III. BOLTZMANN SUPERPOSITION PRINCIPLE

The dynamic mechanical properties of a given polymer, whether represented in the form of storage modulus or loss modulus, can be utilized to predict the behavior of the same polymer in a different flow field. This is done by appealing to the *Boltzmann superposition principle*, whose essence can be understood with the help of the following argument [23].

If a viscoelastic material is subjected to a small, instantaneous strain γ at time zero, a stress τ arises, and this relaxes over a period of time. At any later time t, we have

$$\tau(t) = G(t)\gamma \tag{6.5}$$

in which $G(t)$ is called the stress relaxation modulus. Furthermore, if a series of small strains $\gamma_1, \gamma_2, \ldots$, is imposed on the material at times $t_1, t_2 \ldots$, in the past, the stress at the present time t is a linear combination of the stresses resulting from each of these strains; i.e.,

$$\tau(t) = G(t - t_1)\gamma_1 + G(t - t_2)\gamma_2 + \cdots \tag{6.6}$$

where $(t - t_i)$ is the time elapsed since the imposition of strain γ_i. Converting the sum in Eq. (6.6) to an integral gives

$$\tau(t) = \int_{-\infty}^{t} G(t - s)\,d\gamma = \int_{-\infty}^{t} G(t - s)\dot{\gamma}\,ds \tag{6.7}$$

where s is a past time, $\dot{\gamma}$ is the rate of deformation, and $G(t - s)$ is the yet-to-be-determined stress relaxation modulus. Equation (6.7) can be generalized to three dimensions to give the Boltzmann superposition principle:

$$\tau_{ij}(t) = \int_{-\infty}^{t} G(t - s)\dot{\gamma}_{ij}\,ds \tag{6.8}$$

with $\dot{\gamma}_{ij}$ being the components of the rate of deformation tensor.

The Boltzmann superposition principle is the embodiment of the theory of linear viscoelasticity, and it is valid for both steady and transient deformations, provided the extent of deformation is small. To apply it to a small-amplitude sinusoidal deformation with $\gamma = \gamma_0 \sin \omega s$, we first change the independent variable in Eq. (6.7) from s to t', where $t' = t - s$. Thus,

$$\tau(t) = \int_{0}^{\infty} G(t')\dot{\gamma}\,(t - t')\,dt' \tag{6.9}$$

$$\dot{\gamma} = \omega\gamma_0 \cos\left[\omega(t - t')\right] \tag{6.10}$$

Inserting Eq. (6.10) into Eq. (6.9) and using a trigonometric identity, we get

$$\frac{\tau(t)}{\gamma_0} = \omega \cos(\omega t) \int_0^\infty G(t')\cos(\omega t')dt' + \omega \sin(\omega t) \int_0^\infty G(t')\sin(\omega t')dt' \tag{6.11}$$

which when compared to Eq. (6.4) yields the following expressions for the storage and loss moduli, respectively:

$$G'(\omega) = \omega \int_0^\infty G(t')\sin(\omega t')dt' \tag{6.12}$$

$$G''(\omega) = \omega \int_0^\infty G(t')\cos(\omega t')dt' \tag{6.13}$$

Since the storage and loss moduli are derived from the same stress relaxation modulus, they are not independent of each other. Equations (6.12) and (6.13) allow us to obtain the limiting behavior of the two moduli at very low and very high frequencies. The results are [23]:

$$\lim_{\omega \to 0} G'(\omega) = \left[\int_0^\infty t'G(t')dt'\right]\omega^2 \tag{6.14}$$

$$\lim_{\omega \to 0} G''(\omega) = \left[\int_0^\infty G(t')dt'\right]\omega \tag{6.15}$$

$$\lim_{\omega \to \infty} G'(\omega) = G(0) \tag{6.16}$$

$$\lim_{\omega \to \infty} G''(\omega) = 0 \tag{6.17}$$

Equations (6.14) and (6.15) reveal that double logarithmic plots at low frequencies should have a slope of 2 for the storage modulus and a slope of unity for the loss modulus. This is shown in Figure 6.3a. However, we may not always observe these limits due to the inability to measure small stresses accurately at very low frequencies.

In a steady-shearing flow at a constant shear rate $\dot{\gamma}$, the Boltzmann superposition principle [Eq. (6.9)] predicts that

$$\tau = \dot{\gamma} \int_0^\infty G(t')dt' \tag{6.18}$$

Since the ratio of the shear stress to the shear rate is the viscosity, a combination of Eqs. (6.15) and (6.18) say that

$$\eta_0 = \lim_{\omega \to 0} \frac{G''(\omega)}{\omega} \tag{6.19}$$

and the viscosity appearing in Eq. (6.19) is the zero-shear-rate viscosity, because the theory is valid for small deformations only. This requires that the product of deformation rate with time be small, and since time can be large, the deformation rate must be small. Recall that Eq. (6.19) was seen earlier as Eq. (3.6). In Chapter 3, it was also mentioned that the zero-shear viscosity was independent of the polydispersity index and depended uniquely on the weight-average molecular weight. This is possible only if the low-frequency limit of G'' is a unique function of $\bar{M}_w$; such is, indeed, found to be the case [17].

In closing this section, we mention that linear viscoelasticity theory [Eq. (6.8)] always gives a zero first normal stress difference in shear flow. However, Eq. (5.1) relates the low-frequency limit of the storage modulus to the first normal stress difference in steady laminar shearing flow. This relationship is based on the predictions of the second-order approximation to a simple fluid with fading memory [28], and it has been confirmed by experimental results obtained by a very large number of authors [15].

IV. STRESS RELAXATION MODULUS

In order to predict the stress response of a linear viscoelastic material in a specific flow field through the use of Eq. (6.9), we need the stress relaxation modulus $G(t')$. This may be obtained entirely by experiment or by a combination of experiment and theoretical models, and, in the latter case, mechanical analogs have proven to be particularly popular. Specifically, we consider the stress in a polymer to be composed of an elastic contribution and a viscous contribution, and, in the simplest situation, we represent the polymer as a spring and *dashpot* in series. This is shown schematically in Figure 6.8. The spring is Hookean, for which the stress and strain are related by

$$\tau = g\gamma \tag{6.20}$$

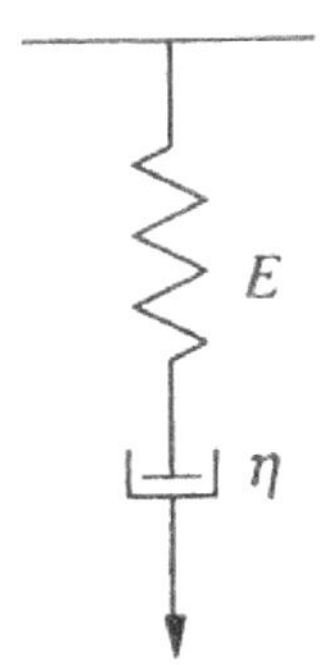

FIGURE 6.8 A Maxwell element. (From Ref. [9].)

whereas the dashpot is Newtonian, so

$$\tau = \eta \frac{d\gamma}{dt} \tag{6.21}$$

Here, g is the spring modulus and η is the dashpot viscosity; the linear combination of the spring and dashpot is called a *Maxwell element.*

It is clear that at any time, the total strain γ in the Maxwell element is a sum of the individual strains in the spring and dashpot, i.e.,

$$\gamma = \gamma_s + \gamma_d \tag{6.22}$$

where the subscripts s and d denote the spring and the dashpot, respectively. Differentiating Eq. (6.22) with respect to time yields

$$\frac{d\gamma}{dt} = \frac{d\gamma_s}{dt} + \frac{d\gamma_d}{dt} \tag{6.23}$$

Because the applied stress τ equals the stress in both the spring and the dashpot, the terms on the right side of Eq. (6.23) can be obtained from Eqs. (6.20) and (6.21), with the following result

$$\frac{d\gamma}{dt} = \frac{1}{g}\frac{d\tau}{dt} + \frac{\tau}{\eta} \tag{6.24}$$

If the polymer is initially stress-free, i.e., $\tau = 0$ at $t = -\infty$, then the solution of Eq. (6.24) is

$$\tau(t) = g\int_{-\infty}^{t} e^{-(t-s)/\theta}\left(\frac{d\gamma}{ds}\right)ds \tag{6.25}$$

in which s is a dummy variable of integration and θ equals η/g. Comparing Eqs. (6.8) and (6.25), we find that

$$G(t-s) = g\exp\left[-\frac{(t-s)}{\theta}\right] \tag{6.26}$$

which when introduced into Eq. (6.5), reveals that

$$\tau(t) = g\gamma e^{-t/\theta} \tag{6.27}$$

and it is seen that in a stress relaxation experiment, the stress decays over a time scale of the order of θ. Consequently, θ is called a *relaxation time*. Although a single relaxation time might fit data on monodisperse polymers, it will certainly not suffice for polydisperse materials. It is for this reason that, even for monodisperse samples, we modify the mechanical analog to make it consist of N Maxwell elements in parallel; each spring has a modulus g_i, and each dashpot has a damping constant η_i. The stress relaxation modulus now becomes

$$G(t-s)=\sum_{i=1}^{N} g_i \exp\left[-\frac{(t-s)}{\theta_i}\right] \tag{6.28}$$

in which $\theta_i = \eta_i / g_i$ and the set of constants g_i and θ_i is called a discrete relaxation time spectrum.

In principle, the relaxation time spectrum can be obtained by fitting Eq. (6.28) to stress relaxation data. In practice, however, it is easier to measure the storage and loss moduli and to use these data to compute the relaxation time spectrum. This computation requires that we be able to interrelate one linear viscoelastic function to another, and available methods have been discussed by Ferry [1]. Here we employ the technique proposed by Baumgaertel and Winter [29], because this is particularly straightforward. If we introduce the stress relaxation modulus, Eq. (6.28), into Eqs. (6.12) and (6.13) and carry out the integrations, we get

$$G'(\omega)=\sum_{i=1}^{N} \frac{g_i(\omega\theta_i)^2}{1+(\omega\theta_i)^2} \tag{6.29}$$

$$G''(\omega)=\sum_{i=1}^{N} \frac{g_i(\omega\theta_i)}{1+(\omega\theta_i)^2} \tag{6.30}$$

The constants g_i and θ_i are found by simply fitting Eqs. (6.29) and (6.30) to measured G' and G'' data using a nonlinear least-squares procedure; typically, we choose between one and two relaxation modes per decade of frequency [29]. Figure 6.9 shows a master curve of G' and G'' values in a temperature range of 130–250°C on an injection molding–grade sample of polystyrene; all the data have been combined by means of a horizontal shift using time–temperature superposition with a 150°C reference temperature. The stress relaxation modulus calculated in the manner of Baumgaertel and Winter is displayed in Figure 6.10 along with actual stress relaxation data at 150°C; the agreement is very good.

So far, we have considered a *discrete spectrum of relaxation times*, i.e., the stress relaxation modulus was made up of a finite number of terms that traced their origin to an equal number of Maxwell elements in the mechanical analog of the polymer. By letting the number of Maxwell elements become infinite, we can define a *continuous relaxation time spectrum* $H(\theta)$ and say that $H\, d(\ln\theta)$ is the contribution to

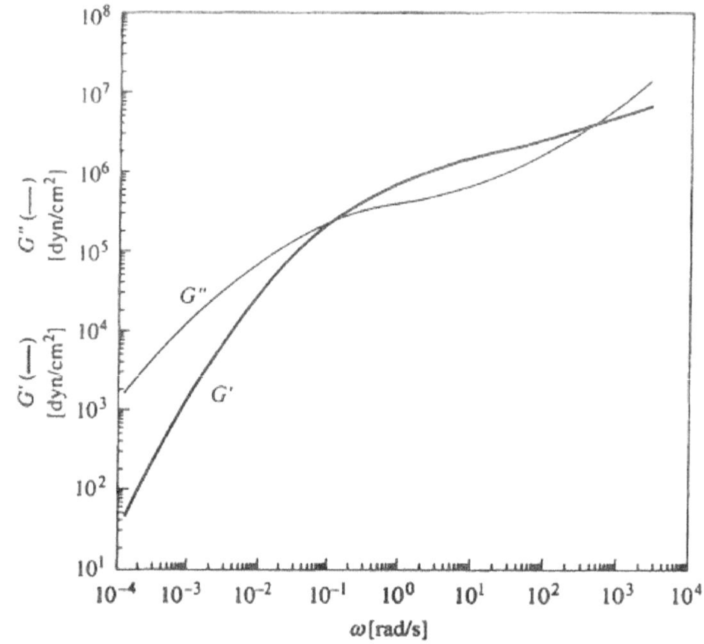

FIGURE 6.9 Master curve of storage and loss moduli of polystyrene. (From Ref. [23].)

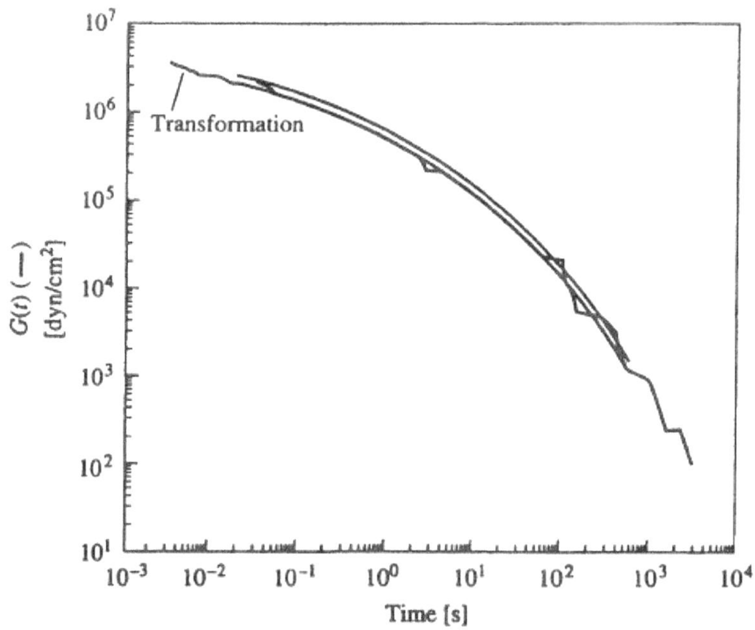

FIGURE 6.10 Stress relaxation modulus of polystyrene at 150°C. (From Ref. [23].)

the modulus from relaxation times that range in value from θ to $\theta + d\theta$. The stress relaxation modulus now takes the form [30]

$$G(t) = \int_{-\infty}^{\infty} H(\theta)\left[\exp\frac{-t}{\theta}\right] d(\ln\theta) \tag{6.31}$$

and dynamic mechanical data can again be used to compute $H(\theta)$: details are available in the literature [1, 30–33].

We re-emphasize that once the stress relaxation modulus has been determined, whether in the form of Eq. (6.28) or Eq. (6.31), the Boltzmann superposition principle [Eq. (6.8)], can be employed to calculate the stress resulting from any specified deformation, provided that the total deformation is small or the rate of deformation is slow, so that linear behavior is exhibited.

Dynamic mechanical measurements are made on solid-in-liquid suspensions and on liquid–liquid emulsions as well, and these are discussed in later chapters. Here, we simply mention that the use of small amplitude oscillatory shear (SAOS) flow is particularly useful for probing the structure and morphology of polymer nanocomposites, i.e., suspensions where the solids are nanoparticles, which are defined as particles where at least one dimension is in the nanometer range. These have a very large surface area to volume ratio, and, as a result, surface forces become dominant, leading to the formation of a three-dimensional network structure at relatively low solids loadings. It is found that there are rheological changes upon crossing the percolation threshold, and these are visible in the storage and loss moduli [34].

V. DYNAMIC RHEOLOGICAL PROPERTIES OF SOLUTIONS

Typical dynamic rheological behavior of polymer solutions at several concentrations is shown in Figures 6.11a and 6.11b [35, 36]. Both the storage modulus and the loss modulus increase with the concentration of polymer. At these low concentrations, few if any molecular entanglements are to be expected, but the solutions still show elasticity. This elasticity, no doubt, results from the distortion of the coiled molecules by the shearing field, with resultant orientation of some of the molecular segments. When the flow stops, the deformed molecules relax back to their normal, randomly coiled state. In concentrated solutions, entanglements can exist that greatly enhance the elasticity. Consequently, for high-molecular-weight polymers, the storage modulus, at a fixed frequency, is a highly nonlinear function of concentration. The changes in the moduli, though, are much less at high frequencies than at low frequencies when the concentration is varied. Also, it is found that data taken at different constant temperatures and a fixed concentration obey the principle of time–temperature superposition [35, 37].

It is common practice to represent the behavior of unentangled polymer solutions with the help of bead-spring theories (see Refs. [1, 23, 38, 39] and Section III of Chapter 4). Here each polymer molecule is modeled as a chain of $N + 1$ beads connected by N elastic springs. By making simplifying assumptions, such as the absence of any polymer–solvent interactions, we can carry out a force balance and determine the tension in each spring resulting from flow-induced extension and orientation

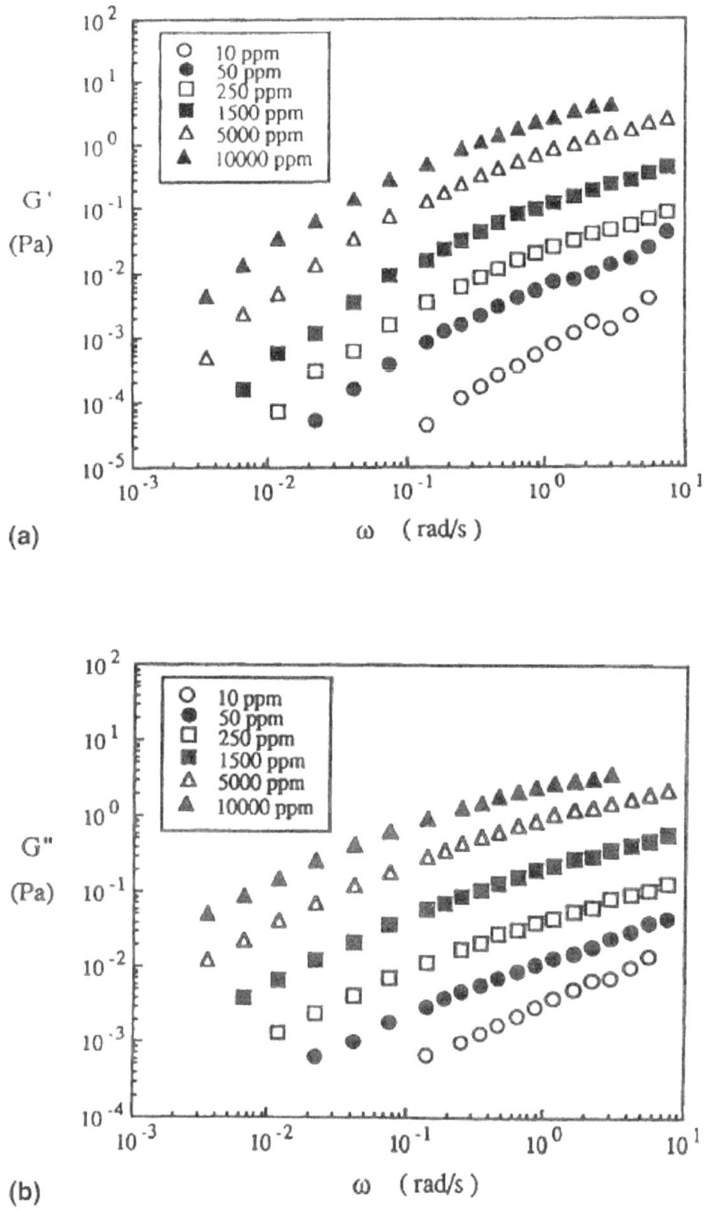

FIGURE 6.11 Dynamic data for solutions of Separan AP30 polyacrylamide in distilled water in a concentration range of 10–10,000 ppm: (a) storage modulus versus frequency, (b) loss modulus versus frequency. (From Ref. [36]. Copyright CRC Press, Boca Raton, Florida.)

of the molecule. The stress contributed by the polymer is then obtained by adding together the contributions from all the different molecules. In the Rouse theory, it is assumed that the molecule is free draining, i.e., that flow is not affected by the presence of the polymer. The resulting expressions for the storage and loss moduli are [38]:

$$G'(\omega) = \frac{cRT}{M} \sum_{i=1}^{N} \frac{\omega^2 \theta_i^2}{1+\omega^2\theta_i^2} \tag{6.32}$$

$$G''(\omega) = \omega\eta_s + \frac{cRT}{M} \sum_{i=1}^{N} \frac{\omega\theta_i}{1+\omega^2\theta_i^2} \tag{6.33}$$

in which c is the mass concentration of polymer, R is the universal gas constant, T is the absolute temperature, M is the polymer molecular weight, and the relaxation times θ_i are given by

$$\theta_i = \frac{6(\eta_0 - \eta_s)M}{\pi^2 i^2 cRT} \tag{6.34}$$

wherein η_0 is the solution zero-shear viscosity and η_s is the solvent viscosity.

In the *Zimm theory*, hydrodynamic interactions are taken into account, but this does not change the functional form of G' and G''; only the magnitude of the various relaxation times changes [38]. In terms of data representation, therefore, Eqs. (6.32) and (6.33) suggest that we plot results in dimensionless form as reduced moduli versus reduced frequency, defined by [1]:

$$G'_R = \left(\frac{M}{cRT}\right) G' \tag{6.35}$$

$$(G'' - \omega\eta_s)_R = \left(\frac{M}{cRT}\right)(G'' - \omega\eta_s) \tag{6.36}$$

$$\omega_R = \left[\frac{(\eta_0 - \eta_s)M}{cRT}\right]\omega \tag{6.37}$$

If the theory is correct, this form of plotting the storage and loss moduli ought to lead to concentration-and-temperature-independent master curves. This is, indeed, found to be the case for aqueous polyacrylamide solutions, and results are displayed in Figure 6.12 [37]; the theory is also shown to apply to aqueous solutions of xanthan gum and carboxymethyl cellulose, used in food applications [37]. As expected, the reduced storage modulus has a slope of 2 at low frequencies, and the reduced loss modulus has a slope of unity. At high frequencies, the Rouse theory predicts a merger of the two curves, each with a slope of 1/2, whereas the Zimm theory says that the two curves become parallel, with a slope of 2/3. It appears that the data of Tam and Tiu [35, 37] follow the trend predicted by the Zimm model.

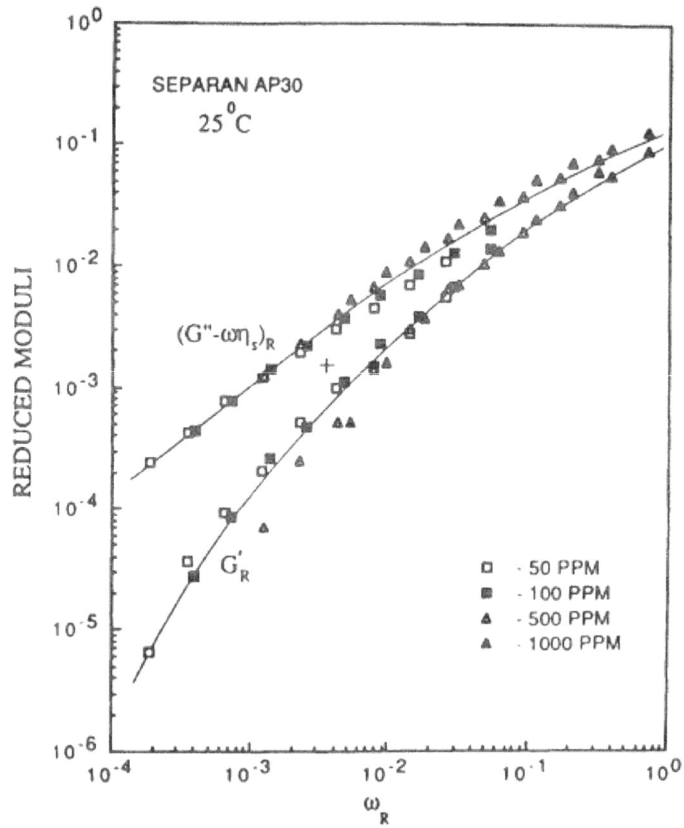

FIGURE 6.12 Reduced dynamic data of four different concentrations of Separan AP30 (polyacrylamide) shown in Figure 6.11. (From Ref. [37].)

In closing this section, we mention that Osaki [40] has examined the viscoelastic properties of extremely dilute polymer solutions and used the collected data to carry out a detailed investigation of the appropriateness of different molecular theoretical explanations.

VI. RELEVANCE TO NONLINEAR VISCOELASTICITY

As mentioned earlier, the Boltzmann superposition principle is the correct constitutive equation for viscoelastic liquids provided that the total strain is small. Indeed, as Bird et al. demonstrate [41, p. 283], serious errors can arise if this theory is used to predict stresses for large deformations. The problem is related to the fact that the infinitesimal strain tensor appearing in Eq. (6.8) is not valid for large deformations, and a finite measure of strain is required. Now, however, there is a great choice in admissible strain measures that all reduce to the same quantity for infinitesimal deformations [30, 41]; picking a specific strain measure is usually justified by comparison with experimental data, and the *Finger measure of strain* has proven particularly popular [30]. With the use of this strain measure, nonlinear behavior can be described, but predicted stress values are usually higher than those measured experimentally. The problem appears to be related to the fact that the stress relaxation modulus becomes a function of strain for large strains. For many polymers, however, the nonlinear relaxation modulus $G(t, \gamma)$ can be written as a product of the linear relaxation modulus $G(t)$ and a function of strain, $h(\gamma)$, which is called the *damping function* and whose value is less than unity [42, 43]. Different functional forms of the damping function result in different stress predictions, and the "correct" form is still to be decided. As a consequence, the development and testing of constitutive theories of nonlinear viscoelasticity is still an active area of research.

While the theoretical relationship between linear and nonlinear viscoelasticity is a matter of debate, for many polymers, certain empirical relationships have been found to be true between material functions characteristic of small strains and those characteristic of large strains. In particular, the modulus of the complex viscosity η^*, defined as

$$|\eta^*| = \left[(\eta')^2 + (\eta'')^2\right]^{1/2} \tag{6.38}$$

where the dynamic viscosity $\eta' = G''/\omega$ and $\eta'' = G'/\omega$, when plotted versus frequency, often superposes with the steady-shear viscosity as a function of the shear rate [46]. This is known as the Cox-Merz rule, and it provides information about a nonlinear property from a measurement of a linear property. Figure 6.13 shows

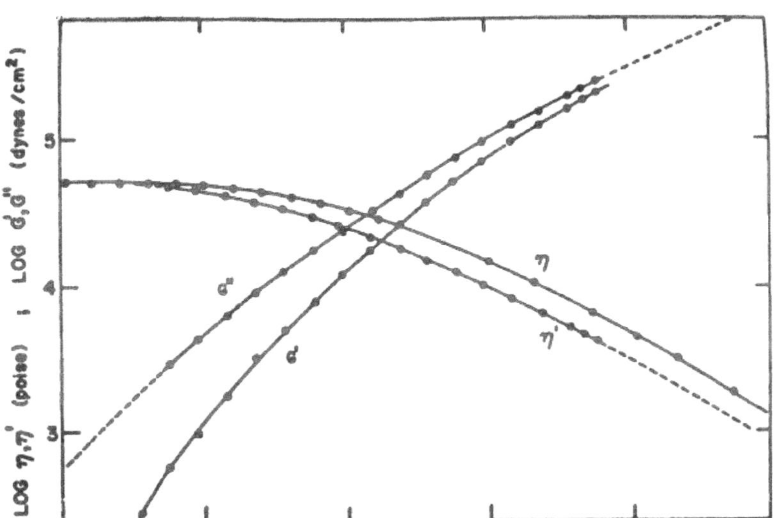

FIGURE 6.13 Steady-shear viscosity, dynamic viscosity, storage modulus, and loss modulus as a function of either angular frequency or shear rate. The absolute value of the complex viscosity is nearly identical to the shear viscosity. (From Ref. [44].)

typical steady-shear and dynamic properties for a low-density polyethylene melt [44], while the steady-shear viscosity equals the dynamic viscosity only at low deformation rates, the magnitude of the complex viscosity is essentially the same as the shear viscosity over the complete frequency or shear rate range. This behavior has been observed by others as well, especially for low-density polyethylene [47].

Another useful empiricism, Laun's rule, relates the first normal stress coefficient to dynamic data [45]:

$$\psi_1 = \frac{2G'}{\omega^2}\left[1+\left(\frac{G'}{G''}\right)^2\right]^{0.7}_{\omega=\dot{\gamma}} \tag{6.39}$$

Figure 6.14 shows that for melts of polyethylene, polypropylene, and polystyrene, Eq. (6.39) does a good job of interrelating linear and nonlinear viscoelastic data.

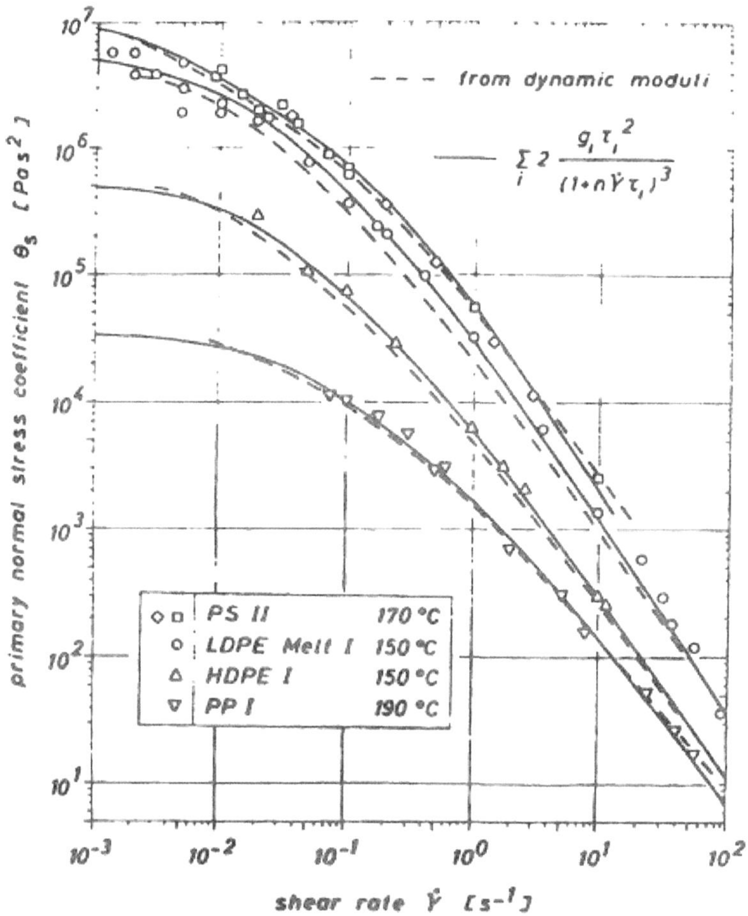

FIGURE 6.14 Primary normal stress coefficient versus shear rate. Symbols represent experimental data. Broken lines are the predictions of Eq. (6.39). (From Ref. [45].)

VII. LARGE AMPLITUDE OSCILLATORY SHEAR

If one examines the behavior of the dynamic moduli at a constant frequency, one finds that, beyond a certain strain amplitude, the stress response becomes non-sinusoidal. However, it must be ensured that there is no slip at the fixture wall and there is no edge fracture of the sample. An indication of the change in the stress response is the fact that both the storage modulus and the loss modulus, computed according to Eq. (6.4), begin to decrease with increasing γ_0. Below this strain, one has *SAOS*, and this strain value represents the limit of linear viscoelasticity. Beyond this point, G' and G'' provided by the instrument lose physical significance. When τ is plotted as a function of γ or $\dot{\gamma}$, where $\dot{\gamma} = \dot{\gamma}_0\, cos\, \omega t$, an ellipse will arise. This is validated by experimental results, provided one is in the linear region (Figure 6.15a). However, the ellipse gets distorted once the limit of linear viscoelasticity is breached (Figure 6.15b). Nonetheless, it is becoming common to make measurements in the non-linear region, and the results represent material response to *large amplitude oscillatory shear* (LAOS) [48]; the distorted ellipses are known as *Lissajous-Bowditch plots.*

In this flow situation, one finds that the stress wave now contains higher harmonics, i.e.,

$$\tau(t) = \sum_{m=1,odd}^{n} \tau_m \sin\left(m\omega t + \delta_m\right) \tag{6.40}$$

in which τ_m and δ_m both depend on γ_0 and ω [48]. Equation (6.40) can also be rewritten as,

$$\tau(t) = \gamma_0 \sum_{m=1,odd}^{n} \left[G'_m\left(\omega\gamma_0\right) sin\, m\omega t + G''_m\left(\omega\gamma_0\right) cos\, m\omega t\right] \tag{6.41}$$

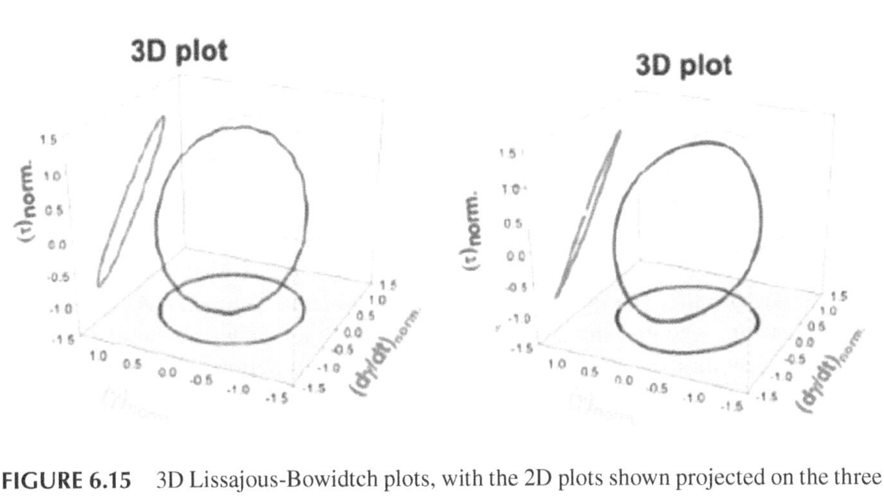

FIGURE 6.15 3D Lissajous-Bowidtch plots, with the 2D plots shown projected on the three planes. (a) Linear region plot; (b) Nonlinear region plot. (From Ref. [53].)

Here, G'_m and G''_m, both depending on γ_0 and ω, are amplitudes of the nth harmonics with frequency $m\omega$, and one recognizes that in the limit of $\gamma_0 \to 0$, G'_1 becomes the storage modulus G', while G''_1 becomes the loss modulus G'' of SAOS.

The advantage of LAOS is that the strain and the strain rate can be varied independently. The disadvantages are the added mathematical complexity, physical interpretation, and difficulty in making accurate measurements. Note, though, that the maximum value of n in Eq. (6.40) is rarely taken to be more than seven [49]. As far as measurements are concerned, these can be made with the help of a cone-and-plate or a parallel plate device, although sample fracturing can occur with polymer melts at a total deformation of a few strain units. Possible remedies include roughened surfaces, small gaps, and the use of special geometries, such as a cone and a partitioned plate [49]. Giacomin and Dealy recommend the use of a sliding-plate viscometer fitted with a flush-mounted, shear stress transducer [48]. Additional information is available in earlier reviews on the subject of LAOS [50, 51].

LAOS data have practical as well as theoretical value. While small amplitude results perturb the molecules only slightly, the imposition of large amplitudes provides information under non-equilibrium conditions, and this is useful for polymer processing. In other words, conducting LAOS experiments is no different than measuring the three possible material functions in steady shear or the tensile stress in constant stretch-rate extension. This allows for fingerprinting the characteristics of the material under study, and these can be related to differences in molecular structure. Note that two materials may behave similarly in SAOS but differently in LAOS. One can also use the results for subjecting rheological constitutive equations to rigorous testing [52]. How this might be done is explained in the review by Kamkar et al. [53].

One begins by defining the intensity I_m of each harmonic as:

$$I_m = \sqrt{(I'_m)^2 + (I''_m)^2} \tag{6.42}$$

in which m takes odd values, $I'_m = \gamma_0 G'_m$ and $I''_m = \gamma_0 G''_m$, and these are obtained by experiment. One then examines I_3/I_1 designated as $I_{3/1}$. The different regions of the strain can be defined as follows: $I_{3/1} \approx 0$ for SAOS, $10^{-4} < I_{3/1} < 10^{-2}$ for MAOS, and $I_{3/1} > 10^{-2}$ for LAOS (here, MAOS is the *medium amplitude oscillatory shear*), and one finds that this quantity is negligible in the linear region but increases with the second power of γ_0 in the nonlinear region for linear, monodisperse, and comb polymers melts. This allows the definition of a nonlinear parameter, $\boldsymbol{Q} = (\boldsymbol{I}_{3/1}) / \boldsymbol{\gamma}_0^2$; Q is a function of strain and frequency. At low strains, Q is a constant, and its value as $\gamma_0 \to 0$ is termed Q_0; it is independent of amplitude but dependent on frequency. When Q_0 is plotted as a function of $\lambda\omega$, the result is a straight line of slope 2 up to $\lambda\omega \approx 1$. Here λ is the fluid relaxation time. This behavior is predicted by most viscoelastic equations. In the higher frequency region, however, predictions depend on the model used, and the predicted slope can range from −1 to 1. Expressions for Q_0 as a function of $\lambda\omega$ are given in [53] for a large number of established constitutive equations; comparison with measurements on a fluid of interest can suggest which fluid model is likely to be the most applicable to the fluid under consideration.

An alternative analysis method known as the *Sequence of Physical Processes* (SPP) seeks to provide a quantitative framework for physical interpretations of LAOS data [54]. Using the Frenet–Serret theorem from differential geometry, instantaneous storage and loss moduli, G'_t and G''_t, respectively, are defined as

$$G'_t = \frac{d\sigma}{d\gamma} \tag{6.43}$$

$$G''_t = \omega \frac{d\sigma}{d\dot{\gamma}} \tag{6.44}$$

These moduli, having been calculated at every point on the 3D Lissajous-Bowditch closed loop curve (Figure 6.15), are typically visualized on plots of G''_t versus G'_t and time derivatives of G''_t versus G'_t, typically known as *Cole-Cole plots*. The results are then used to interpret and describe the structural evolution of the material from the linear to the nonlinear region [51, 53, 55, 56].

In summary, to obtain physical understanding of the material behavior during LAOS, a number of different methods of analysis have been developed [51, 53]. Measures obtained from the different analyses are being employed to explain structural changes of the material in the nonlinear region [51, 53], and results from the different analyses are being used to explain physically the structure of complex soft materials such as foods [55, 56]. It is worth noting that LAOS analysis of a range of food materials has become popular and is being explored for a wide range of applications—from structural changes during processing to textural perception during consumption [53, 55–59]. Due to the smaller linear viscoelastic region observed with most foods, a deeper understanding of the physical and structural changes when subject to LAOS tests would be practically useful.

VIII. PRACTICAL APPLICATIONS

This chapter has been concerned mainly with the theoretical aspects of dynamic mechanical analysis, and these apply to both synthetic and natural polymers. The relevance of these measurements, however, extends to a large number of practical situations as well. For practical applications, the SAOS measurements are more conveniently performed as a function of temperature, at a fixed strain (Figure 6.2). It is less cumbersome and time-consuming than the time–temperature superposition position technique required to obtain results such as those shown in Figure 6.3a, and conveniently applies to thermorheologically complex materials, a trait of many industrial formulations, such as foods. The sensitivity of the temperature dependence of the dynamic mechanical properties to a number of different factors, such as glassy and crystalline transitions, molecular structure relaxations, degree of crosslinking, compatibility of blends of polymers and low-molecular-weight components, fillers, etc., makes it a powerful technique for practical applications such as characterizing and optimizing formulations (see Chapter 4 in [60]) and guide processing. A simplified schematic illustrating the effect of different factors is shown in Figure 6.16.

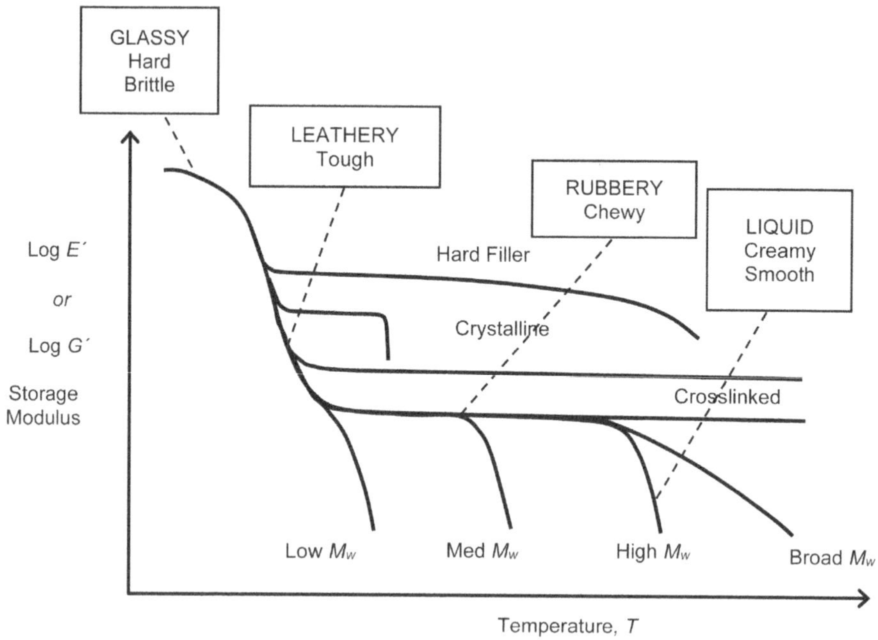

FIGURE 6.16 Effect of different factors on the temperature dependence of G'. (From Ref. [61].)

The formulator may begin with the choice of the base polymer and add other polymers and additives to modify the properties, so the final formulation serves the application optimally. A pressure-sensitive adhesive, typically consists of a blend of an elastomer and a low-molecular-weight resin. The bond strength measured by the mechanical failure test is commonly observed to be a product of the surface thermodynamics and the adhesive rheology [62]. Through the choice of the resin structure, molecular weight, and concentration, a wide range of rheological behavior is achieved due to the degree of compatibility between the resins and the natural rubber and styrene-butadiene elastomers [63–65]; this in turn can lead to different performance from the adhesives. Figure 6.17 illustrates how the polymer properties are altered when blended with a low-molecular-weight resin.

While we consider foods to be made up of fats, polysaccharides, and proteins, a processed food product is composed of multiple components. The interactions amongst these components can be revealed from the temperature dependence of the dynamic mechanical properties, where one examines the storage and/or loss modulus as a function of temperature at a fixed frequency. An example is displayed in Figure 6.18 for a wheat dough.

Wheat dough can be considered to be a suspension of starch granules in an aqueous mixed polymer matrix. The structural transitions and transformations that happen when dough is heated from a very low temperature are mostly irreversible, and these are identified in Figure 6.18. This information is very helpful in delineating the processing window for bakery products.

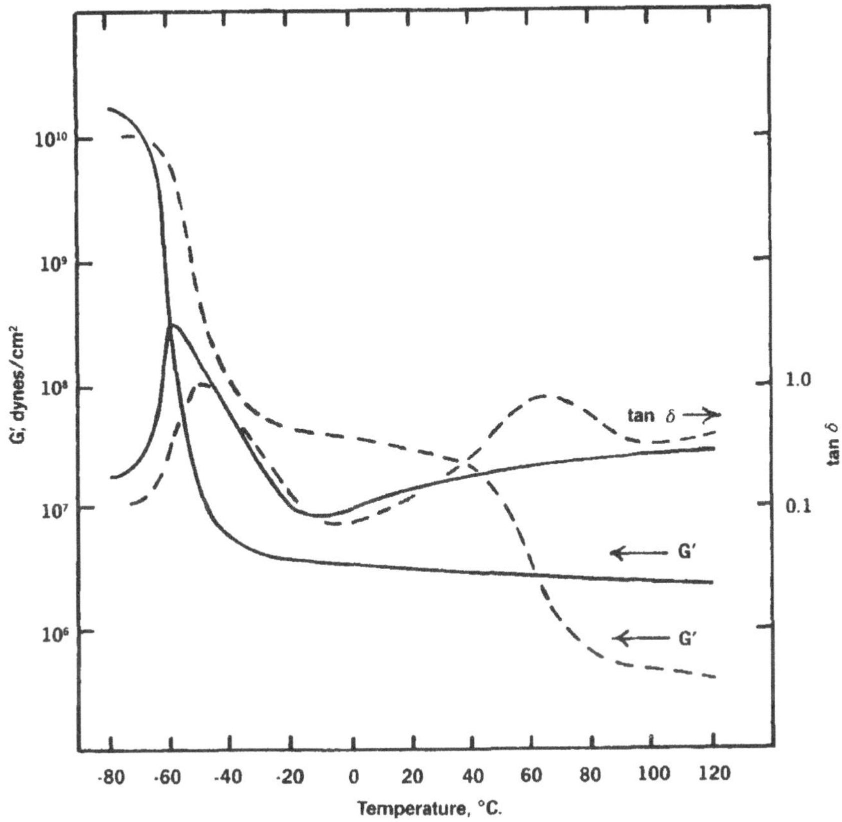

FIGURE 6.17 G' and tan δ versus temperature for a 1:1 blend of natural rubber with polystyrene resin: (—) NR (- - -) NR:PS. (From Ref. [63].)

Figure 6.19 shows the temperature dependence of the storage modulus of a hard candy and two different formulations of soft candy. Hard candy is formulated with low-molecular-weight components, and it shows a progressive decrease in the storage modulus with increasing temperature. Soft candy, on the other hand, contains macromolecules, and it not only displays a rubbery plateau characteristic of polymeric materials, it also reveals a variety of other transitions. These can be related to processability and texture during consumption.

In recent years, there has been a push to reduce fat in the diet, and Figure 6.20 shows that large differences arise in the rheology of cream cheese as the fat content is reduced, and these may alter the consumer experience. This trend toward healthier foods is likely to dominate research on processed foods in the future, and it may involve the elimination of some ingredients and the simultaneous addition of some nutrients. There has also been a shift from animal-based proteins to plant-based proteins. Regardless of whether this trend continues or not, rheological procedures, in general, and dynamic mechanical techniques, in particular, will be helpful in designing foods.

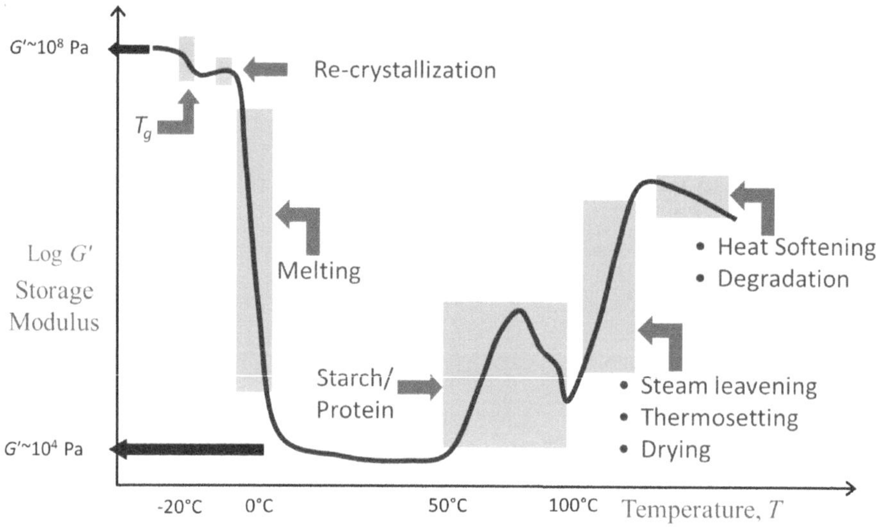

FIGURE 6.18 Rheological thermal analysis of wheat dough.

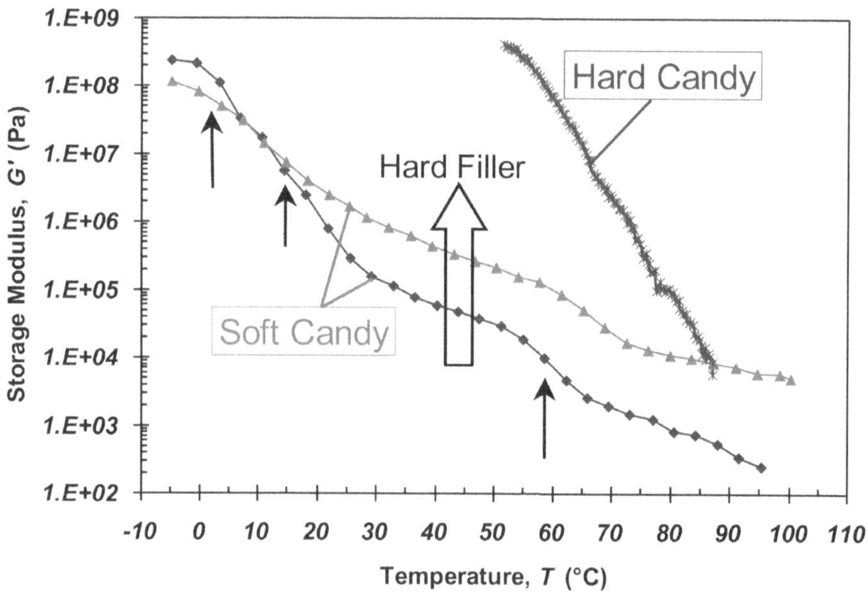

FIGURE 6.19 Rheological thermal analysis of three different candy formulations. (From Ref. [61].)

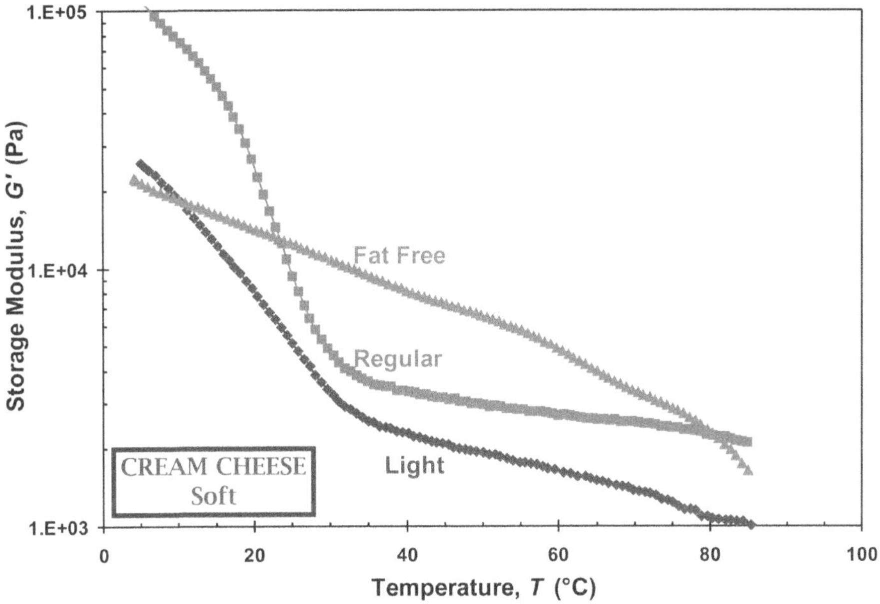

FIGURE 6.20 Influence of fat content on the rheological thermal analysis of cream cheese. (Data courtesy of Lauriane Scanu (personal communication).)

IX. CONCLUDING REMARKS

The major thermal transition observed with polymers, whether amorphous or crystalline, is the glass-to-rubber transition or T_g. However, other transitions are also observed at temperatures lower than T_g, and these can again be identified as peaks in the plot of tan δ versus temperature. These are called secondary dispersions, and they can be related to the motion of side chains or still smaller groups attached to the main chain. Dynamic data also reveal information on polymer crystallinity, chain orientation, and the possibility of plasticization due to changes in humidity levels or the diffusion of gases [3].

We also mention that we have focused on dynamic shear moduli in this chapter. We could equally well have worked with compliances, but the information content is the same because the complex modulus can easily be converted to the complex compliance.

The treatment described in this chapter has also been extended to multicomponent and composite systems. Here the dynamic properties depend not just on the properties of the constituent materials but also on how they are put together and the nature of any joints and interfaces. Thus, results can be used to determine the effectiveness of any adhesives used. This work has been motivated by aircraft and aerospace applications, among others [3].

Dynamic mechanical properties of polymer melts and polymer solutions are relatively easy to measure, and these are measured on a routine basis for polymer and food characterization purposes. Often these measurements are directly related to the shear viscosity or the first normal stress coefficient in shear. They also allow for the determination of the relaxation time spectrum, which, together with the Boltzmann superposition principle, gives information on any small-strain property of interest. As we shall see later, dynamic testing is also extremely useful for probing the structure and examining the behavior of crystalline polymers, crosslinking systems, solid-in-liquid suspensions, emulsions, polymer blends, and polymer composites. A particular advantage of small-amplitude deformations is that material structure is not altered during the course of measurement; this is especially important for multiphase systems. By contrast, LAOS flow does alter material structure and is representative of non-equilibrium structures, and it has, to a large extent, been pursued mainly in academic research laboratories. Measuring reliable LAOS data is more difficult, and their analysis is highly mathematical; many methods of analysis have been published, with consensus not having been reached on the most useful method.

The temperature dependence of the dynamic mechanical properties is easier to perform and more useful in practice. It is sensitive to a number of different factors and thus useful to study structural changes, transitions, and transformations due to the presence of different additives and processing.

REFERENCES

1. J. D. Ferry. Viscoelastic Properties of Polymers. 3rd ed., Wiley, New York, 1980.
2. L. E. Nielsen. Polymer Rheology. Marcel Dekker, New York, 1977.
3. T. Murayama. Dynamic Mechanical Analysis of Polymeric Material. Elsevier, Amsterdam, 1978.
4. J. M. Dealy. Rheometers for Molten Plastics. Van Nostrand Reinhold, New York, 1982.
5. R. W. Whorlow. Rheological Techniques. 2nd ed., Ellis Horwood, New York, 1992.
6. G. Marin. Oscillatory rheometry. In: A. A. Collyer and D. W. Clegg, eds. Rheological Measurement. 2nd ed., Chapman & Hall, London, 1998, pp. 3–46.
7. B. Maxwell and R. P. Chartoff. A polymer melt in an orthogonal rheometer. Trans. Soc. Rheol. 9:41–52 (1965).
8. R. K. Krishnaswamy and D. S. Kalika. Dynamic mechanical relaxation properties of poly(ether ether ketone). Polymer 35:1157–1165 (1994).
9. R. K. Gupta. Polymer and Composite Rheology. 2nd ed., Marcel Dekker, New York, 2000.
10. J. E. Mark, A. Eisenberg, W. W. Graessley, L. Mandelkem, and J. L. Koenig. Physical Properties of Polymers. American Chemical Society, Washington, DC, 1984, pp. 97–153.
11. W. E. Rochefort and S. Middleman. Rheology of xanthan gum: Salt, temperature and strain effectsin oscillatory and steady shear experiments. J. Rheol. 31:337–369 (1987).
12. G. R. Zeichner and P. D. Patel. Proc. 2nd World Congr. Chem. Eng. Vol. 6, p. 373, Montreal, 1981.
13. P. A. M. Steeman. A numerical study of various rheological polydispersity measures. Rheol. Acta. 37:583–592 (1998).
14. M. Ansari, S. G. Hatzikiriakos, A. M. Sukhadia, and D. C. Rohlfing. Rheology of Ziegler-Natta and metallocene high-density polyethylenes: Broad molecular weight distribution effects. Rheol. Acta. 50:17–27 (2011).
15. S. Onogi, T. Masuda, and K. Kitagawa. Rheological properties of anionic polystyrenes. I. Dynamic viscoelasticity of narrow-distribution polystyrenes. Macromolecules. 3:109–116 (1970).
16. L. R. G. Treloar. The Physics of Rubber Elasticity. 3rd ed., Clarendon, Oxford, 1975.

17. T. Masuda, K. Kitagawa, T. Lnoue, and S. Onogi. Rheological properties of anionic polystyrenes. II. Dynamic viscoelasticity of blends of narrow-distribution polystyrenes. Macromolecules. 3:116–125 (1970).
18. T. Masuda, Y. Ohta, and S. Onogi. Rheological properties of anionic polystyrenes. III. Characterization and rheological properties of four-branch polystyrenes. Macromolecules. 4:763–768 (1971).
19. A. Mendelson. Flow properties of polyethylene melts. Polym. Eng. Sci. 9:350–355 (1969).
20. D. Yan, W.-J. Wang, and S. Zhu, Effect of long chain branching on rheological properties of metallocene polyethylene. Polymer. 40:1737–1744 (1999).
21. W. H. Tuminello. Molecular weight and molecular weight distribution from dynamic measurements of polymer melts. Polym. Eng. Sci. 26:1339–1347 (1986).
22. D. W. Mead. Determination of molecular weight distributions of linear flexible polymers from linear viscoelastic material functions. J. Rheol. 38:1797–1827 (1994).
23. A. Kumar and R. K. Gupta. Fundamentals of Polymer Engineering. 3rd ed., CRC Press, Boca Raton, FL, 2019.
24. S. Wu. Dynamic rheology and molecular weight distribution of insoluble polymers: Tetrafluoroethylene-hexafluoropropylene copolymers. Macromolecules. 18:2023–2030 (1985).
25. C. D. Han and M. S. Jhon. Correlations for the first normal stress difference with shear stress and of the storage modulus with loss modulus for homopolymers. J. Appl. Polym. Sci. 32:3809–3840 (1986).
26. C. D. Han and H.-K. Chuang. Criteria for rheological compatibility of polymer blends. J. Appl. Polym. Sci. 30:4431–4454 (1985).
27. S. G. Hatzikiriakos. Long chain branching and polydispersity effects on the rheological properties of polyethylenes. Polym. Eng. Sci. 40:2279–2287 (2000).
28. G. Astarita and G. Marrucci. Principles of Non-Newtonian Fluid Mechanics. McGraw Hill, London, 1974.
29. M. Baumgaertel and H. H. Winter. Determination of discrete relaxation and retardation time spectra from dynamic mechanical data. Rheol. Acta. 28:511–519 (1989).
30. J. M. Dealy and K. F. Wissbrun. Melt Rheology and Its Role in Plastics Processing. Van Nostrand Reinhold, New York, 1990.
31. R. I. Tanner. Note on the iterative calculation of relaxation spectra. J. Appl. Polym. Sci. 12:1649–1652 (1968).
32. J. Honerkamp and J. Weese. A nonlinear regularization method for the calculation of relaxation spectra. Rheol. Acta. 32:65–73 (1993).
33. H. H. Winter. Analysis of dynamic mechanical data: inversion into a relaxation time spectrum and consistency check. J. Non-Newt. Fluid Mech. 68:225–239 (1997).
34. S. Bhattacharya, R. K. Gupta, and S. Bhattacharya. The rheology of polymeric nanocomposites. In: R. K. Gupta, E. Kennel, and K.-J. Kim, eds. Polymer Nanocomposites Handbook. CRC Press, Boca Raton, FL, 2010, pp. 151–203.
35. K. C. Tam. Rheology and turbulent flow behavior of polymer solutions. PhD dissertation, Monash University, Australia, 1990.
36. K. C. Tam and C. Tiu. Water soluble polymers (rheological properties). In: J. C. Salamone, ed. Polymeric Materials Encyclopedia, Vol. 11, CRC Press, Boca Raton, FL, 1996, pp. 8655–8677.
37. K. C. Tam and C. Tiu. Steady and dynamic shear properties of aqueous polymer solutions. J. Rheol. 33:257–280 (1989).
38. W. W. Graessley. The entanglement concept in polymer rheology. Adv. Polym. Sci. 16:1–179 (1974).
39. R. B. Bird, C. F. Curtiss, R. C. Armstrong, and O. Hassager. Dynamics of Polymeric Liquids. 2nd ed. Vol. 2, Wiley, New York, 1987.

40. K. Osaki. Viscoelastic properties of dilute polymer solutions. Adv. Polym. Sci. 12:1–64 (1973).
41. R. B. Bird, R. C. Armstrong, and O. Hassager. Dynamics of Polymeric Liquids. 2nd ed. Vol. 1, Wiley, New York, 1987.
42. M. H. Wagner. Analysis of time-dependent non-linear stress-growth data for shear and elongational flow of a low-density branched polyethylene melt. Rheol. Acta. 15:136–142 (1976).
43. V. H. Rolon-Garrido and M. H. Wagner. The damping function in rheology. Rheol. Acta. 48:245–284 (2009).
44. R. N. Shroff. Dynamic mechanical properties of polyethylene melts: Calculation of relaxation spectrum from loss modulus. Trans. Soc. Rheol. 15:163–175 (1971).
45. H. M. Laun. Prediction of elastic strains of polymer melts in shear and elongation. J. Rheol. 30:459–501 (1986).
46. W. P. Cox and E. H. Merz. Correlation of dynamic and steady flow viscosities. J. Polym. Sci. 28:619–622 (1958).
47. M. Padmanabhan and M. Bhattacharya. In-line measurements of rheological properties of polymer melts. Rheol. Acta. 33:71–87 (1994).
48. A. J. Giacomin and J. M. Dealy. Using large-amplitude oscillatory shear. In: A. A. Collyer and D. W. Clegg, eds. Rheological Measurement. 2nd ed., Chapman & Hall, London, 1998, pp. 327–356.
49. G. W. Kamykowski. Large amplitude oscillatory shear for characterizing polymers – A tutorial. Paper presented at Society of Plastics Engineers ANTEC, Las Vegas, NV, April 29, 2014.
50. R. H. Ewoldt, A. E. Hosoi, and G. H. McKinley. New measures for characterizing nonlinear viscoelasticity in large amplitude oscillatory shear. J. Rheol. 52:1427–1458 (2008).
51. K. Hyun, M. Wilhelm, C. O. Klein, K. S. Cho, J. G. Nam, K. H. Ahn, S. J. Lee, R. H. Ewoldt, and G. H. McKinley. A review of nonlinear oscillatory shear tests: Analysis and application of large amplitude oscillatory shear (LAOS). Prog. Polym. Sci. 36:1697–1753 (2011).
52. B. Debbaut and H. Burhin. Large amplitude oscillatory shear and Fourier-transform rheology for a high-density polyethylene: Experiments and numerical simulation. J. Rheol. 46:1155–1176 (2002).
53. M. Kamkar, R. Salehiyan, T. B. Goudoulas, M. Abbasi, C. Saengow, E. Erfanian, S. Sadeghi, G. Natale, S. A. Rogers, A. J. Giacomin, and U. Sundararaj. Large amplitude oscillatory shear flow: Microstructural assessment of polymeric systems. Prog. Polym. Sci. 132:101580 (2022).
54. S. A. Rogers. In search of physical meaning: Defining transient parameters for nonlinear viscoelasticity. Rheol. Acta. 56:501–525 (2017).
55. A. J. Sandoval, M. Fernández, O. Sanz, A. Santamaria, E. Penott-Chang, and A. J. Müller. Large amplitude oscillatory shear (LAOS) behavior of chocolates of different compositions. J. Rheol. 66:859–879 (2022).
56. M. Y. Erturk, S. A. Rogers, and J. Kokini. Comparison of sequence of physical processes (SPP) and Fourier transform coupled with Chebyshev polynomials (FTC) methods to interpret large amplitude oscillatory shear (LAOS) response of viscoelastic doughs and pectin solutions. Food Hydrocoll. 128:107558 (2022).
57. P. Ptaszek. Large amplitude oscillatory shear (LAOS) measurement and Fourier-Transform rheology: Application to food. In: J. Ahmed, P. Ptaszek, and S. Basu, eds. Advances in Food Rheology and its Applications. Elsevier, New York, 2017, pp. 87–123.
58. H. S. Joyner. Nonlinear (large amplitude oscillatory shear) rheological properties and their impact on food processing and quality. Ann. Rev. Food Sci. Technol. 12:591–609 (2021).

59. M. Y. Erturk, A. N. M. Le., and J. Kokini. Advances in large amplitude oscillatory shear rheology of food materials. Front. Food Sci. Technol. 3:1130165 (2023).
60. L. E. Nielsen and R. F. Landel. Mechanical Properties of Polymers and Composites. 2nd ed., Marcel Dekker, New York, 1994.
61. M. Padmanabhan. The application of rheological thermal analysis to foods. Proc. 3rd Int. Symp. Food Rheol. Struct., Zurich, pp. 447–448, February 9–13, 2003.
62. M. Tirrell. Current methods and issues in adhesion science. In: J. W. Mintmire, R. J. Colton, P. P. Schmidt, and R. G. Brandt, eds. Workshop on Adhesion and Intermolecular Forces. Naval Research Laboratory, Report Number NRL/MR/6179-92-7143 (1992).
63. J. B. Class and S. G. Chu. The viscoelastic properties of rubber-resin blends. I. The effect of resin structure. J. Appl. Polym. Sci. 30:805–814 (1985).
64. J. B. Class and S. G. Chu. The viscoelastic properties of rubber-resin blends. II. The effect of resin molecular weight. J. Appl. Polym. Sci. 30:815–824 (1985).
65. J. B. Class and S. G. Chu. The viscoelastic properties of rubber-resin blends. III. The effect of resin concentration. J. Appl. Polym. Sci. 30:825–842 (1985).

7 Extensional Viscosity

I. INTRODUCTION

As pointed out in Chapter 1, polymeric liquids and foods can undergo tensile deformation in addition to the more familiar shear deformation, and its importance has been highlighted in several books, book chapters, and technical articles [1–4 and references therein]. This is due to both practical and theoretical reasons: (1) industrial polymer fabrication processes such as fiber spinning, film blowing, blow molding of bottles, and flat film extrusion involve a predominantly tensile deformation, and this mode of deformation is also encountered in the peel testing of pressure-sensitive adhesives [5]. (2) Similarly, food processing operations such as sheeting, extrusion puffing, and baking, as well as during the steps involved during eating, impart extensional flows ([4] and Appendix). (3) Due to the fact that polymer molecules uncoil and stretch during extensional flow, the resistance to deformation in extension is significantly greater than the corresponding resistance to shear flow. Consequently, the *uniaxial extensional viscosity* of non-Newtonian liquids, especially at large values of the stretch rate, is hundreds of times greater than the shear viscosity, and, in general, one quantity cannot be obtained from a measurement of the other. Note that the ratio of the extensional viscosity to the zero-shear viscosity is generally called *Trouton's ratio:* for Newtonian liquids, Trouton's ratio has a value of 3, in uniaxial extension. When the steady shear viscosity is dependent on shear rate, one may define the Trouton ratio as the ratio of the extensional viscosity to the shear viscosity, but the shear viscosity is evaluated at a shear rate equal to √3 times the stretch rate [6]. This follows from requiring that the second invariant of the rate of deformation tensor be the same in the two flow fields. An important consideration when assessing extensional flows in practical applications is to recognize that there are three different types of extensional flows: *uniaxial*, *biaxial*, and *planar* [6].

If the stretch rate in a uniaxial extension experiment (that begins from rest at time zero) is small, we can use the theory of linear viscoelasticity, Eq. (6.8), to compute the extensional viscosity. Employing the Boltzmann superposition principle in combination with Eqs. (1.14), (1.17), and (1.18), we find that the tensile stress growth coefficient is given by

$$\eta_E^+(t) = 3\int_0^t G(t')dt' = 3\eta^+(t) \tag{7.1}$$

in which η^+ is the *shear stress growth coefficient* defined as the ratio of the instantaneous shear stress to the shear rate during startup of shearing at a constant shear rate. At steady state, of course, the preceding result leads to Eq. (1.15). In other words, the steady-state behavior of polymeric fluids is Newtonian, provided we are in the linear viscoelastic region. The predicted relationship between η_E^+ and η^+ is, indeed, observed,

DOI: 10.1201/9781003013501-7

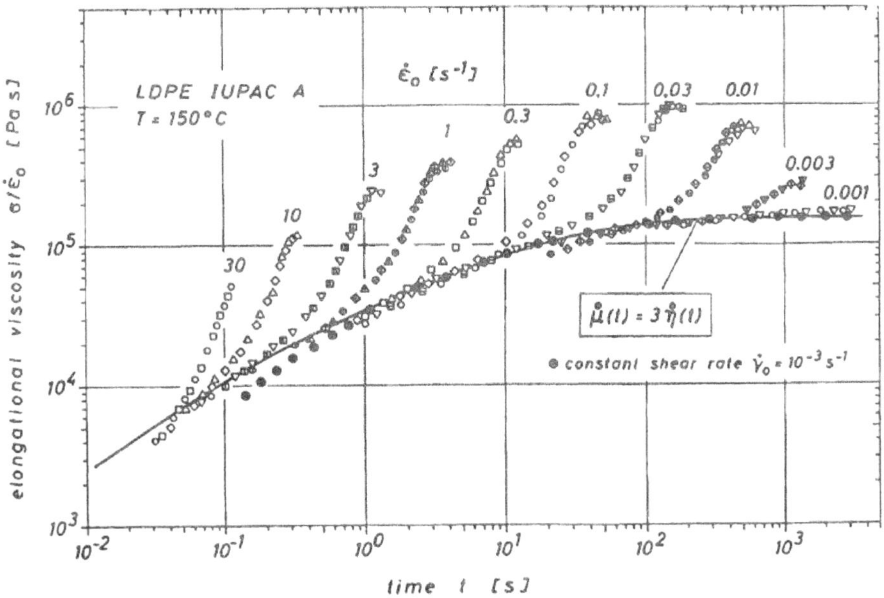

FIGURE 7.1 Tensile stress growth data for LDPE. (From Ref. [7]. Used by permission of Steinkopff Publishers, Darmstadt, FRG.)

and it can be seen to hold at the lowest stretch rate of 0.001 sec^{-1} in Figure 7.1, which shows tensile stress growth data on a low-density polyethylene sample [7]; at higher deformation rates, however, the data deviate from linear behavior, and the deviation occurs at smaller strain values with increasing stretch rates. The occurrence of stresses that are higher than those predicted by the theory of linear viscoelasticity is known as *strain hardening.* Note from Figure 7.1 that a steady state in the stress is not attained at the highest stretch rate values.

Observed steady-state extensional viscosity behavior is compared with shear viscosity in Figure 7.2. Three general types of tensile behavior have been reported in the past as a function of tensile stress (or rate of elongation). Tensile viscosity may be nearly independent of tensile stress for some polymers [8, 9]. In other cases, extensional viscosity starts to increase with tensile stress at a value of stress about the same as where the shear viscosity starts to decrease [9]; this behavior is known as *extension thickening.* In still other cases, particularly those involving suspensions, the elongational viscosity starts to decrease at about the same value of stress as where the shear viscosity decreases from the Newtonian value [9, 10]; this behavior is known as *extension thinning.* Another observation is that one polymer (linear low-density polyethylene (LLDPE)) may have a higher viscosity than another polymer (low-density polyethylene (LDPE)) in shear, but the situation can be reversed in extension [11]. Note though that stress levels or stretch rates in industrial polymer processing operations are usually significantly higher than those indicated in Figure 7.2; however, these flows are typically of short duration, and steady states in the stress are rarely, if ever, attained. Furthermore, tensile viscosity increases with decreasing temperature,

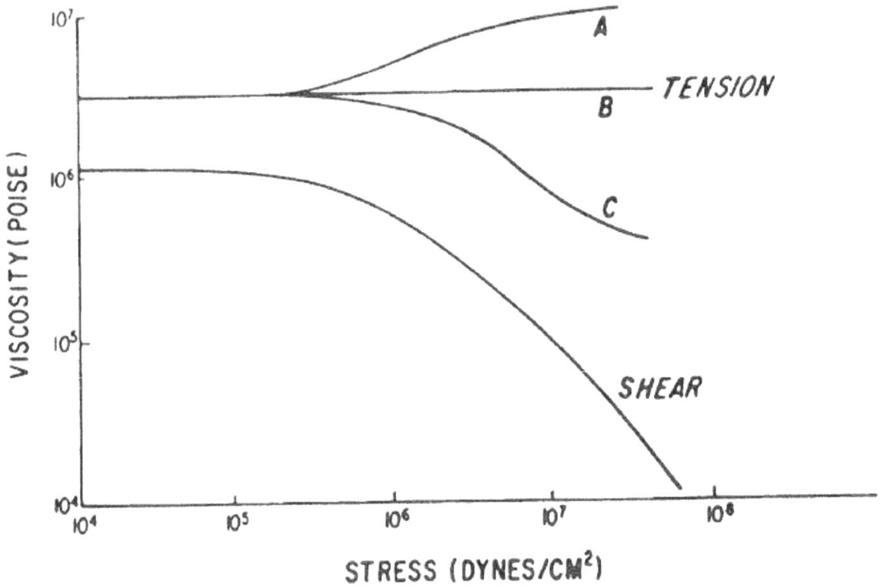

FIGURE 7.2 Three types of tensile viscosity behaviors compared to shear viscosity as a function of tensile or shear stress for polymeric fluids. (From Ref. [12].)

and this increase often overshadows the change in tensile stress due to changes in the rate of deformation. At present, there is no single theory capable of quantitatively predicting this wide diversity of behavior found for extensional viscosity.

In some polymer processing operations, such as fiber spinning, it is desirable that for process stability the extensional viscosity increase with the rate of elongation. The reason for this is as follows: if a weak spot develops in the fiber, which results in a decrease in cross-sectional area, the rate of elongation at that spot increases. The increase in the rate of elongation causes an increase in the tensile viscosity, which resists further stretching of the thin section. In tubular film blowing, it is found that the response of a polymer melt to an extensional deformation, which depends on the structure of the polymer, determines how easily the film thins and how stable it is against rupturing [13]. Thus, extensional viscosity measurements can be useful in selecting polymers that are easier to process. Wheat doughs exhibit strain hardening, and this is considered critical to the expansion of bread dough during baking without premature collapse [Appendix].

Extensional flow measurements can also be made on polymer solutions, and these data are again relevant for a number of industrial applications. It is found, for example, that, unlike Newtonian liquids, drops of viscoelastic liquids suspended in an air stream do not always disintegrate easily. Instead, depending on the tensile stresses generated, they form strands and big blobs, or they form long filaments that gradually break up into a very fine spray [14]. Matta and Tytus [15] showed that the addition of a small amount of polymer to a liquid could change the breakup mechanism from shear to extension, resulting in an order-of-magnitude increase in the average drop size; this phenomenon is clearly important in fuel atomization and the aerial spraying of insecticides. Similarly, in roll coating of paints, high-molecular-weight thickeners modify the surface-tension-induced breakup of roll fibers, and this can lead to increased spatter [16–18].

Biaxial extensional flow behavior is important in many applications to foods, such as in the growth of bubbles during the baking of bread and the expansion of extrudates in extrusion cooking. During the swallowing of foods, the *bolus* is subject to extensional flows [4]. For patients suffering from difficulty in swallowing (a medical condition known as *Dysphagia*), beverages thickened with various polysaccharides are formulated to provide the right rheology. Extensional viscosity has been identified as important in optimizing the formulations due to its role in avoiding the breakup of the bolus during swallowing [19].

II. LOW-STRETCH-RATE BEHAVIOR OF EXTENSIONAL VISCOSITY

In order to understand and interpret extensional flow measurements properly, it is necessary to ensure that the results depend neither on the technique of stretching nor on any assumptions regarding the constitutive behavior of the material; data on dough rheology are presented in this spirit, and their interpretation is deferred to the chapter on suspensions (volume 2). This requires that data be obtained either under constant-stress or under constant-strain-rate conditions. This task has now been accomplished for relatively viscous polymer melts, and commercial instruments are available, but these operate at stretch rates that are significantly lower than those encountered in industrial practice. As far as polymer solutions are concerned, the task is more challenging due to the mobile nature of liquids and the impossibility of gripping them and making them stretch in a prescribed manner. Nonetheless, enough reliable data are now available on many polymer melts and polymer solutions that some general comments can be made regarding the nature of the extensional viscosity. Note that data on unfilled polymer melts and polymer solutions are qualitatively similar, although Trouton's ratio of polymer solutions can exceed Trouton's ratio of polymer melts by about two orders of magnitude [3].

A. Measurement Techniques for Polymer Melts

A conceptually simple way of making extensional viscosity measurements is to clamp or glue one end of a cylindrical sample of molten polymer to a force transducer and to move the other end outwards according to the regimen of Eq. (1.12) [20, 21]. The polymer may be stretched horizontally or vertically, but it is generally necessary to immerse the sample in an inert liquid to prevent gravity from contributing to the deformation. An alternative to controlling the stretch rate is to impose a constant stress and to monitor the resulting strain as a function of time [20, 22–24]; a constant extensional viscosity is achieved if the *Hencky strain* begins to increase linearly with time. Shown in Figure 7.3 is the "creepmeter" developed by Münstedt [24]; for some time, this instrument was commercialized by the Göttfert and Rheometrics companies. A disadvantage of this design is that the duration of the experiment is limited to the time taken by the moving end of the sample to traverse the length of

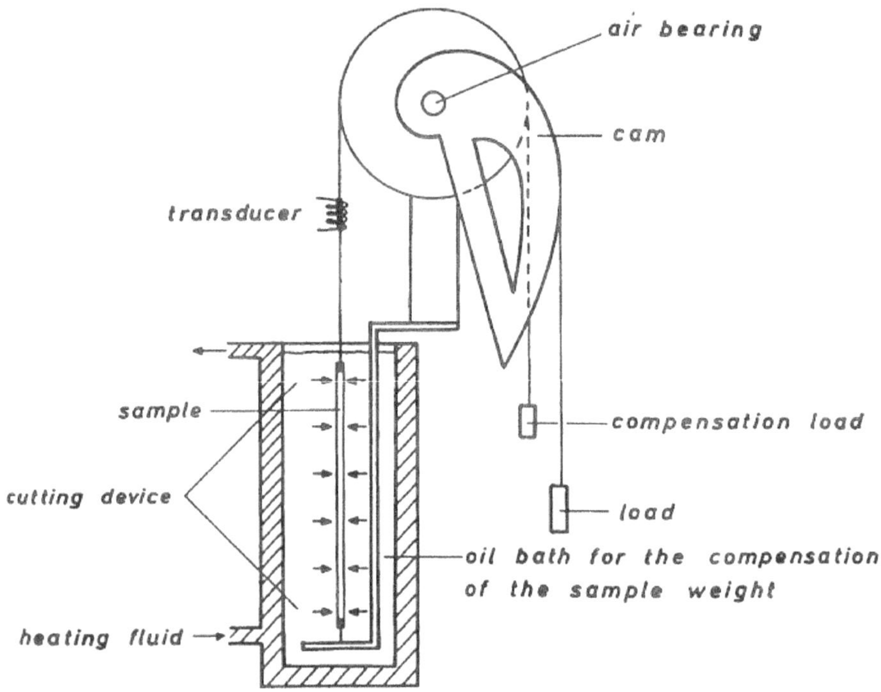

FIGURE 7.3 Schematic drawing of the tensile creep apparatus. (From Ref. [24]. Used by permission of Steinkopff Publishers, Darmstadt, FRG.)

the liquid bath, which typically corresponds to about 3–4 strain units; the upper limit on the stretch rate is a few reciprocal seconds. Note that the same results are obtained regardless of whether a constant strain rate is imposed or whether the stress is held fixed, provided that care is taken to minimize temperature and stretch rate nonuniformities in the sample during the process of deformation. However, the total strain needed to achieve a steady state is found to be smaller if the instrument is operated in the constant-stress mode.

The maximum-strain limitation, imposed by the size of the apparatus, can be removed if one stretches a sample at constant length. This can be done in one of two ways: (1) the two ends are pulled outwards at the same constant speed, or (2) one end is kept stationary while the velocity of the other end is maintained constant. While details of the different designs may be found in the literature [2, 3], the rotary clamp technique of Meissner is particularly noteworthy [25]. Meissner used two pairs of gears to grip and stretch a cylindrical specimen; Hencky strains as large as 7, equivalent to a thousandfold increase in specimen length, could be achieved. Later, Meissner and Hostettler [26] pioneered the development of new clamps that make use of metal conveyor belts for stretching samples that are supported by an inert gas that is heated to the test temperature; the stretching force is monitored with the use of leaf-spring transducers. This design was commercialized by the Rheometrics Scientific Company (now a part of TA Instruments), and a sketch of their elongational rheometer for melts is shown in Figure 7.4. This instrument could stretch samples

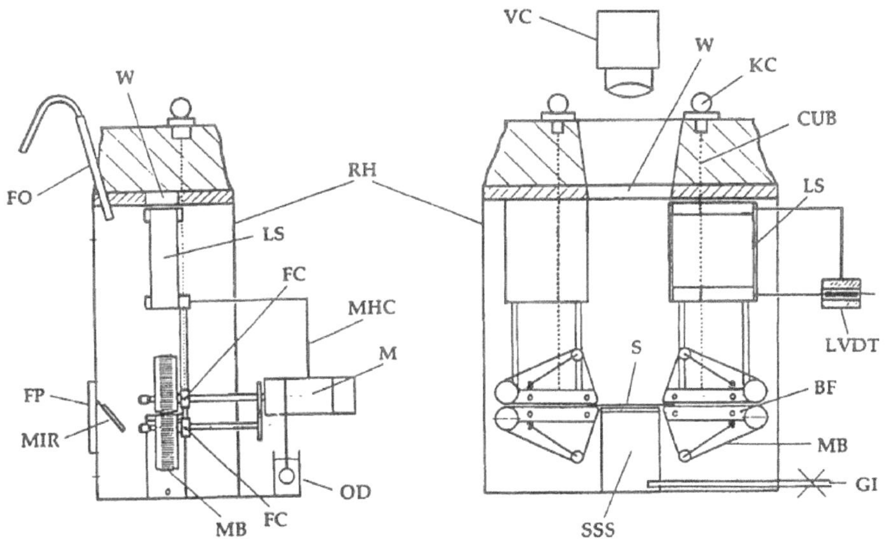

FIGURE 7.4 Rheometrics Scientific elongational rheometer for melts. RH, rheometer housing; MB, metal conveyor belt; BF, belt fixture; S, sample; LS, leaf spring; M, motor; Gl, gas inlet; SSS, sample support system; VC, video camera. (From Ref. [26]. Used by permission of Steinkopff Publishers, Darmstadt, FRG.)

at rates up to 1 sec^{-1}, and it had a maximum attainable strain of 7, operation up to 350°C, and a force measurement range of between 0.001 and 2 N. A horizontal translating clamp design driven by an Instron testing machine was used by de Bruijne et al. [27] to exponentially stretch samples of wheat dough. This permitted the samples to be stretched at constant Hencky strain rates. The samples were floated on mercury to prevent sagging. Padmanabhan [4] demonstrated that their stress–strain curves when replotted as transient uniaxial extensional viscosity exhibit strong strain hardening behavior (Figure 7.5) similar to polymer melts shown in Figure 7.1.

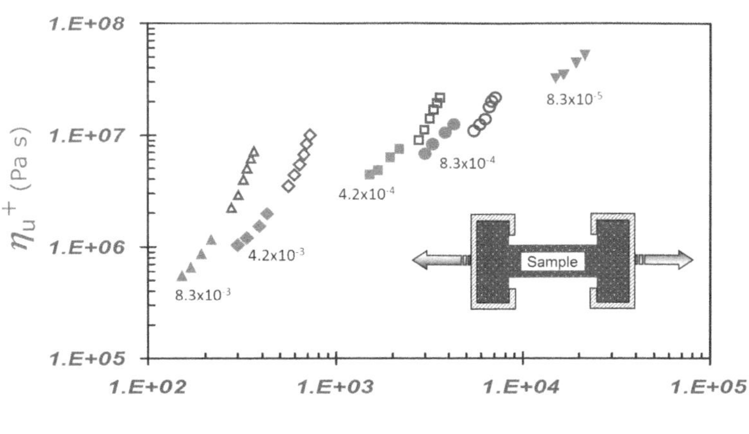

FIGURE 7.5 Transient extensional viscosity of soft and hard wheat doughs. (Figure adapted from Ref. [4].)

Padmanabhan et al. [28] revived the "fiber-windup" technique originally developed by Macosko and Lorntson [29]. One end of the sample was attached to the torque transducer of a commercial rotational rheometer, while the other end was moved outwards by being wound onto a drum. The entire setup was housed inside the oven of the rheometer, and measurements were performed at 150°C. The instantaneous tensile stress was calculated from knowledge of the measured torque and the instantaneous sample radius, and the stretch rate was computed using the drum linear velocity and the length of the sample being stretched. Results for the transient extensional viscosity agreed well with published results on a similar LDPE [30]. Sentmanat further developed the "fiber-windup" technique by designing a detachable fixture for use with commercial rotational rheometers [31, 32]. Known as the *Sentmanat Extensional Rheometer (SER)*, it is shown in Figure 7.6. It is now available as an option on most commercial rheometers. It uses a rectangular strip of polymer having a shear viscosity of at least 10,000 Pa-s and a mass of less than 1 g. The sample, having a length L_0, is attached to two vertical drums, each of radius R, that rotate in opposite directions and stretch the sample. If the angular velocity of the drums is ω, the constant stretch rate imposed on the polymer is $(2\omega R)/L_0$. The stretching force is obtained from the torque exerted on one of the drums; the stress is given by the ratio of this force to the instantaneous sample cross-sectional

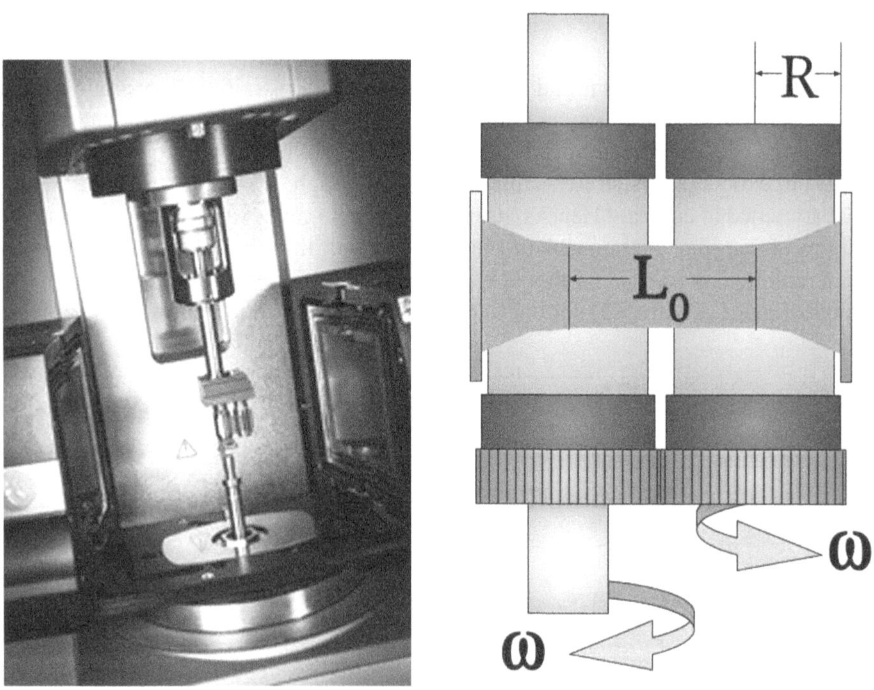

FIGURE 7.6 (a) A picture of the SER mounted on the rheometer (courtesy Abhishek Shetty of Anton Paar from Document B86IA012EN-B- "Sentmanat Extensional Rheometer SER3", Anton Paar Application Report). (b) Schematic of the SER.

area. Results obtained with the use of the SER have been validated by comparison with earlier results on a variety of polymers, including LDPE and LLDPE [32]. Measurements conducted on wheat doughs using the SER were comparable to those from a filament stretching rheometer [33].

B. Experimental Results and Their Constitutive Modeling

The first systematic attempt at obtaining consistent rheological data on chemically similar polymer melts was made by the International Union of Pure and Applied Chemistry (IUPAC) working party on the structure and properties of commercial polymers, which in 1975 reported work done on three LDPE samples [34]. One of these well-characterized samples, IUPAC A (no longer available [32]), was subjected to extensive elongational testing by Laun and Münstedt; Figure 7.1 shows some of their results. The elongational viscosity, calculated using the steady-state stress values in Figure 7.1, is displayed in Figure 7.7 as a function of the stretch rate [35]. The results span six decades of stretch rate and show that Trouton's ratio equals 3 at low stretch rates. As the stretch rate is increased, the extensional viscosity increases, goes through a maximum, and finally decreases to a value even below that of the zero-shear value. At high stretch rates, the slope of the steady extensional viscosity versus stretch rate curve is observed to be −0.5 [32]. (Note that there has been some concern in the Literature about whether or not a true steady state is actually achieved in experiments such as these [36], and the maximum in the transient extensional viscosity is often taken to be the steady uniaxial extensional viscosity [32].)

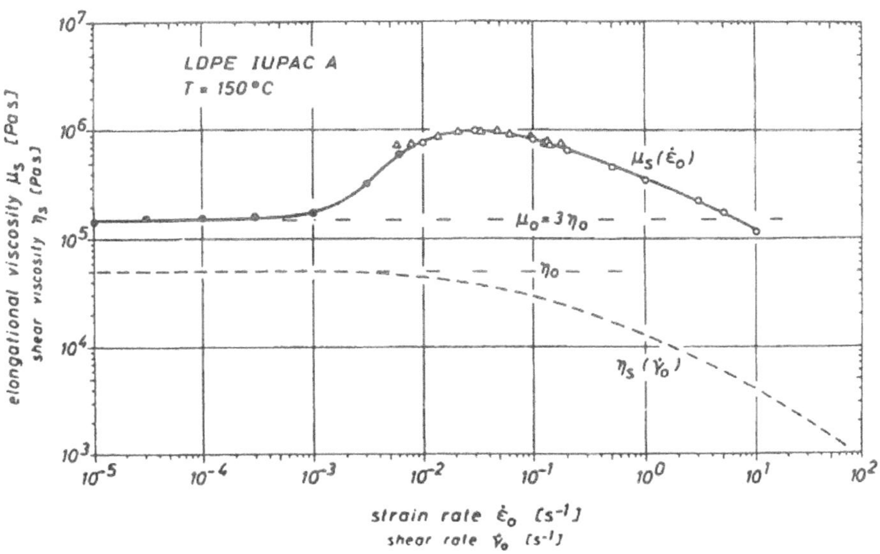

FIGURE 7.7 Shear and elongational viscosities of LDPE as a function of deformation rate calculated using the data of Figure 7.1. (From Ref. [35]. Used by permission of Steinkopff Publishers, Darmstadt, FRG.)

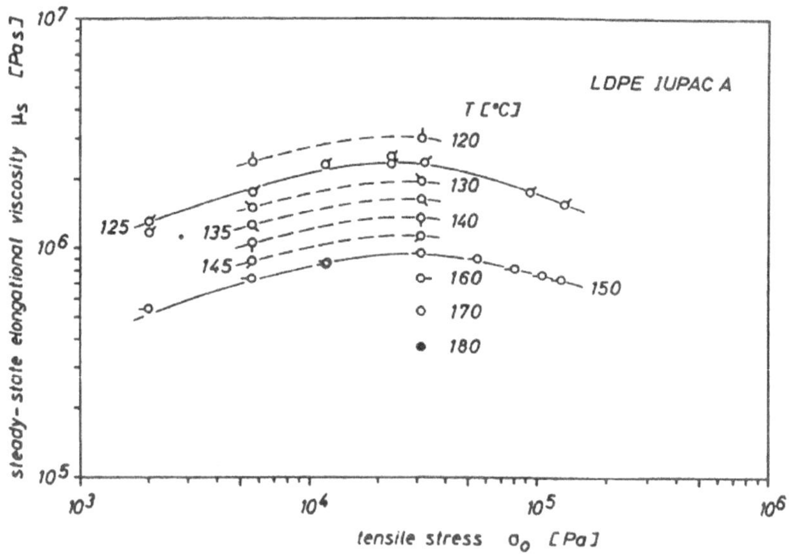

FIGURE 7.8 Stress dependence of steady-state elongational viscosity at different temperatures. (From Ref. [7]. Used by permission of Steinkopff Publishers, Darmstadt, FRG.)

Thermoplastic starch was introduced in Chapter 3 as an important food material. The extensional viscosity of this polymer has been measured, and the behavior appears to be similar to that displayed in Figure 7.1. In each case, there is a linear region, and, depending on the plasticizer used, the linear region is followed by strain hardening above a limiting stretch rate [37].

Münstedt and Laun also determined the temperature dependence of the extensional viscosity of IUPAC A, and their measurements are plotted in Figure 7.8 as a function of tensile stress [7]. It is seen that the shape of the extensional viscosity curve is not altered by changing the temperature. In fact, all the data can be made to overlap by means of a vertical shift, demonstrating the validity of the time–temperature superposition principle; the temperature shift factors are identical to those obtained from shear data.

Variables other than temperature and stretch rate also influence the extensional flow behavior of polymer melts. These are material structural variables such as polymer molecular weight, molecular weight distribution, and chain branching. How these variables influence extensional viscosity appears to depend on whether one is in the linear region or in the nonlinear region. Data on several polystyrenes [38, 39] suggest that, at low deformation rates, the value of the extensional viscosity is a function of the molecular weight alone and is proportional to the 3.4–3.6 power of the weight-average molecular weight. Representative data are shown in Figure 7.9 [38], and these are consistent with the observation that the extensional viscosity of molten polymers is generally equal to three times the shear viscosity at low stretch rates; recall from Chapter 3 that the zero-shear viscosity of entangled polymer melts behaves in a similar manner with changes in the weight-average molecular weight.

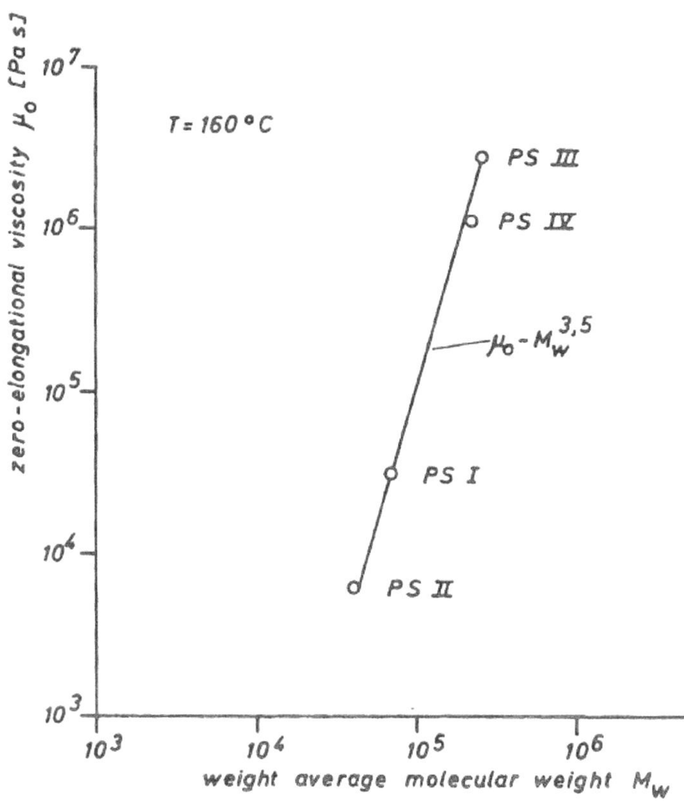

FIGURE 7.9 Molecular weight dependence of elongational viscosity. (From Ref. [38].)

At higher deformation rates, however, the extensional viscosity (unlike the shear viscosity) increases, and it can reach values that are several times larger than the Trouton value with a broadening of the molecular weight distribution. This effect is especially noticeable in the presence of a high-molecular-weight tail, even when the highest-molecular-weight components cannot be detected by the use of a gel permeation chromatograph.

The influence of chain branching on extensional viscosity has been explored in the literature, mainly by comparing and contrasting results obtained using different members of the polyethylene family. Whereas high-density polyethylene (HDPE) is an essentially linear molecule, LDPE and LLDPE are branched, but the branches of LLDPE are relatively short. In qualitative terms, it is found that it is easier to stretch branched polyethylenes in a uniform, homogeneous manner as compared to linear polyethylene [40], although polymers with extremely long branches may be difficult to stretch [41]. In quantitative terms, the presence of long-chain branching (similar to a high-molecular-weight tail) shows up as strain hardening during uniaxial elongation at a constant stretch rate; the longer the branches, the greater is the extent of strain hardening. This is shown in Figure 7.10, where the behavior of LLDPE is compared with that of two LDPEs having different extents of long-chain branching [13]. The weight-average molecular weight of LDPE2 exceeds that of both LDPE1 and LLDPE by at least a factor of 4, and it also has much longer branches as compared to LDPE1. It has the lowest values of tensile stress in the linear viscoelastic region, but with increasing time of stretching, there is immense strain hardening in LDPE2, but very little in LLDPE. If, however, long-chain branching is incorporated into linear copolymers like LLDPE, shear thinning as well as melt elasticity are enhanced [42]. Alternatively, in order to increase the extensional viscosity of LLDPE (whose films are valued for their toughness and tear strength) and to make the process of film blowing more stable, a small amount of LDPE may be blended with LLDPE [13]. Indeed, this is established industrial practice, but the added LDPE does not contribute to film toughness. One way to attain the same objective is to add a free radical generator to LLDPE during processing, and this induces long-chain branching. Using this procedure, Williamson et al. [43] were able to replace a 70% LLDPE + 30% LDPE blend with a 98% LLDPE + 2% LDPE blend and get essentially the same rheology, similar behavior during film blowing, and desirable film mechanical properties. At the other extreme, due to the absence of chain branching, the curve of extensional viscosity versus stretch rate or tensile stress for HDPE shows only a slight maximum; on lowering the molecular weight, the maximum disappears, and the steady-state behavior is Newtonian [44–45]. For all the different-molecular-weight HDPE polymers, though, further increases in the stretch rate result in decreases in the extensional viscosity, which can become significantly lower than thrice the zero-shear viscosity [44]. We caution, though, that exceptions to these generalizations can sometimes be found [46], but these may be due to phase separation resulting from the presence of two different kinds of molecules in the polymer.

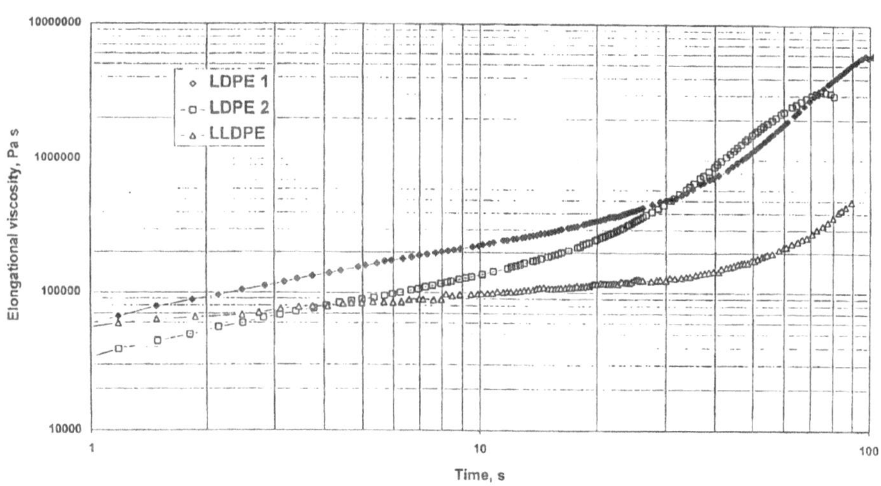

FIGURE 7.10 Tensile stress growth data for three different polyethylenes at 140°C. The stretch rate is 0.052 sec^{-1}. (From Ref. [13].)

The importance of long-chain branching to polymer processing is best illustrated by the behavior of isotactic polypropylene (iPP), a very high-volume, commodity polymer. This is a linear, unbranched molecule that does not strain harden in extensional flow. As a result, it is difficult to thermoform because polypropylene sheets tend to sag during the forming stage. Furthermore, if PP is foamed with the use of a physical blowing agent, there are processing difficulties again, since the low extensional viscosity leads to bubble collapse. However, long-chain branches can be introduced with a variety of methods, including reactive extrusion, the use of catalysts, irradiation, or other chemical means [47–50]. The resulting material is known as "*high melt strength*" polypropylene; the significance of the term "*melt strength*" is explained in Section III.A, and the measurement of melt strength is described there as well. The effect of increasing the branching density on the rheology of PP is seen in Figure 7.11 [51]. In this figure, PP-0 is not irradiated, while PP-10 has the highest radiation dose. For the non-irradiated sample, there is no strain hardening, regardless of the stretch rate imposed, but strain hardening increases as the radiation dose increases. The introduction of long-chain branches clearly changes the extensional flow behavior of PP, and this generally makes foaming easier. It also makes the process of film blowing more stable. We also note that blending some PP having long branches with linear PP makes the blend strain harden, but only at low deformation rates; unlike with polyethylene blends, the strain hardening decreases with increasing stretch rate [52]. Another unusual observation with long chain branched-polypropylene (LCB-PP) is that pre-shearing of the polymer reduces fluid elasticity, in general, and strain hardening, in particular, although the phenomenon,

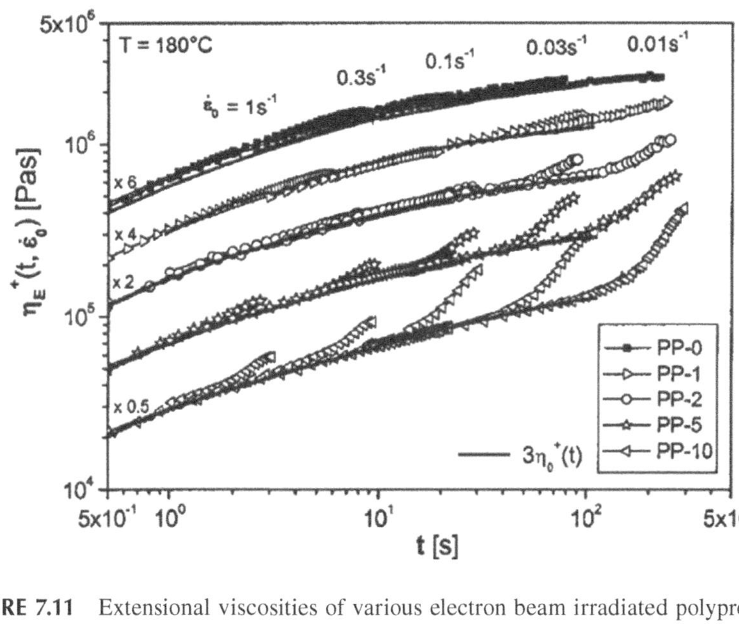

FIGURE 7.11 Extensional viscosities of various electron beam irradiated polypropylenes. (The numbers on the left side of the curves indicate the factors by which the curves were shifted on the y-axis for a clearer distinction.) (From Ref. [51].)

called *shear modification*, is reversible. The cause of these rheological changes is postulated to be either a change in entanglements or alignment of side chains along the backbone of the macromolecule [53].

In terms of the constitutive modeling of data of the kind shown in this and earlier chapters, one wishes that the chosen equation be able to portray key aspects of the observed behavior of polymer melts. These include a shear-thinning viscosity, strain hardening in extension, a large first normal stress difference, and transient effects such as creep and stress relaxation. Depending on the application of the equation, one may also want the equation to describe other phenomena, such as stress overshoots in shear and extension. The simplest approach is to employ a version of the Boltzmann superposition principle, but one that has been generalized to large deformations through the use of a finite measure of strain. As mentioned in Section VI of Chapter 6, the Finger measure of strain C_{ij}^{-1} appears to be the most appropriate one for this purpose. The resulting equation is known as the *Lodge rubber-like liquid* [54, 55]:

$$\tau_{ij} = \frac{g}{\theta} \int_{-\infty}^{t} \exp\left[\frac{-(t-s)}{\theta}\right] C_{ij}^{-1}\, ds \tag{7.2}$$

and its differential counterpart is the *upper-convected Maxwell equation* [55, 56]:

$$\tau_{ij} + \theta \frac{\delta \tau_{ij}}{\delta t} = 2\, g\theta D_{ij} \tag{7.3}$$

in which the D_{ij} are the components of the strain rate tensor and $\delta\tau_{ij}/\delta t$ is the upper-convected derivative given by

$$\frac{\delta \tau_{ij}}{\delta t} = \frac{\partial \tau_{ij}}{\partial t} + \sum_{m=1}^{3} \left[v_m \frac{\partial \tau_{ij}}{\partial x_m} - \tau_{im} \frac{\partial v_j}{\partial x_m} - \tau_{mj} \frac{\partial v_i}{\partial x_m} \right] \tag{7.4}$$

and the two constants g and θ are generally picked to give agreement with linear viscoelastic data; often a discrete relaxation spectrum is used. The Lodge rubber liquid constitutive model, when applied to wheat dough, showed qualitative agreement with experimental data [33].

Although the upper-convected Maxwell equation displays all the qualitative features of nonlinear viscoelastic behavior and it has been used extensively to model polymer processing operations [57], it has some rather severe shortcomings. In particular, this model predicts that the viscosity in shear is constant and the viscosity in extension becomes unbounded at a finite value of the stretch rate. In addition, the second normal stress difference is predicted to be zero. In general, the use of this model results in stresses that are larger than those measured experimentally. Thus, at the very least, an additional parameter is needed to moderate the increase in stress with increasing deformation rate. For integral equations, this additional parameter could be part of the damping function introduced in Section VI of Chapter 6, whereas, for differential models, it could be part of the time derivative. Alternatively, the two constants in Eqs. (7.2) and (7.3) could be made variables, i.e., material functions, and could be made to depend on the invariants of the strain rate tensor. All these possibilities have been proposed, and specific choices have been justified based on physical or mathematical grounds; the resulting modified equations may be found in standard books [56–59]. An especially popular equation is the one due to Phan-Thien and Tanner [60]; when applied to extensional flow data, this equation is often written in the following form [61]:

$$\left[\frac{\alpha}{g}\mathrm{tr}(\tau)\right]\tau_{ij} + \theta\frac{\delta\tau_{ij}}{\delta t} = 2g\theta D_{ij} \tag{7.5}$$

in which α is a new parameter and tr(τ) represents the trace or sum of the diagonal terms of the extra stress tensor.

Khan and Larson [56, 61] examined the ability of several constitutive equations to portray polyethylene melt data in step shear, step biaxial extension, and constant-stretch-rate uniaxial extension. They found that Eq. (7.5) was one of the few simple equations that could fit data in all the three flow fields. However, for the IUPAC A polymer, a different value of the parameter α was required to fit data from each of the three types of deformations. In other words, parameter values that gave a good fit to the data shown in Figure 7.7 did not fit the corresponding shear and biaxial extension data adequately. This kind of situation is encountered frequently in polymer rheology, and one way of addressing this state of affairs is to pick the best parameter values that describe the main aspects of the processing situation of interest. In other words, one must build into the model the answer that one ultimately wishes to obtain!

In recent years, tube models, based on the concept of polymer reptation, have been the subject of intense research. However, these efforts have not resulted in the emergence of an equation that can be recommended for the purpose of simulating polymer processing. In this regard, the reader is referred to the review of tube-model-based constitutive equations by Narimissa and Wagner [62]. It will become clear that this topic is best left to the initiated.

C. Extensional Viscosity of Polymer Solutions

The development of constant-stretch-rate, uniform stretching of polymer solutions lagged behind similar work on polymer melts. However, impetus was provided by a series of international meetings in France between 1988 and 1993. These conferences also led to a large body of published work on the shear and extensional viscoelastic properties of polymer solutions [63], including research on three extremely well-characterized polymer solutions: M1—a constant-viscosity elastic liquid composed of a 0.244% solution of polyisobutylene in a mixed solvent of 7% kerosene in polybutene [64]; A1—a highly elastic, but shear-thinning, liquid made up of 2% polyisobutylene in a mixture of *cis*- and *trans*-decalin [65]; and S1—a shear thinning, viscoelastic solution of 2.5% polyisobutylene dissolved in a mixture of decalin and polybutene and having a relaxation time similar to that of M1 [66]. The participants in these workshop-type meetings used different techniques to make measurements on the same fluid, and not only did this make data comparison easy, but it also made it possible rationally to discuss the merits and demerits of each proposed technique of extensional viscosity measurement.

The technique that ultimately led to the measurement of steady-state extensional viscosity was pioneered by Sridhar and coworkers [67, 68]. These researchers inserted the fluid sample between two identical coaxial disks placed one below the other. The disk diameter ranged from 1 to 7 cm, and the sample thickness was about 2 mm. In the initial experiments [67], the fluid was stretched vertically by keeping the upper disk stationary and moving the lower disk downwards. Figure 7.12 shows a sequence of photographs of the stretching of a constant-viscosity, ideal elastic liquid at an average stretch rate of 0.52 sec^{-1}; except near the disk surfaces, the stretching is fairly uniform, and the filament radius is independent of position over 95% of the filament length [67]. The setup was later refined to work in either the horizontal or the vertical configuration, and a schematic diagram of the apparatus is displayed in Figure 7.13 [68]. In this version, both ends are moved outwards simultaneously, and the maximum final length of the filament is about 1 m; this produces large enough total strains for a steady state in the stress to be reached in many of the experiments. The force is measured using a transducer attached to one of the linear drive units; the filament diameter is determined with a high-resolution optical device. To compensate for end effects, the speed of the disks is continually adjusted to ensure that the filament diameter at the midpoint of the filament decreases exponentially with time, and additional refinements have been made to the apparatus since the time it was

FIGURE 7.12 Sequence of photographs showing the stretching of a filament of a 36 Pa-sec viscosity solution of 0.185% PIB in a solvent of kerosene and polybutene. The average stretch rate is 0.52 sec^{-1}, and the initial diameter and filament length are 3.03 and 1.8 mm, respectively. (From Ref. [67], with permission from Elsevier Science.)

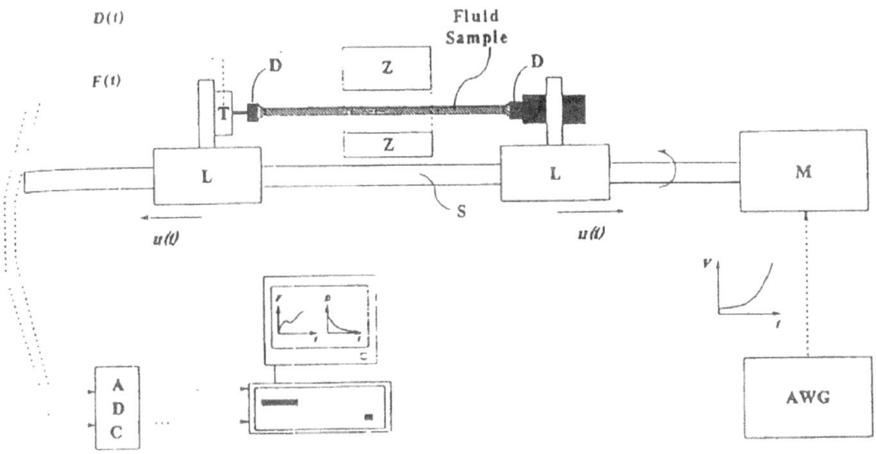

FIGURE 7.13 Schematic diagram of the filament-stretching rheometer: AWG = arbitrary waveform generator, M = servomotor, L = linear drive unit, S = shaft, D = disk, T = force transducer, Z = diameter measuring device, ADC = analog-to-digital converter. (From Ref. [68].)

initially constructed. A version of this instrument, called the Filament Stretching Extensional Rheometer, was available commercially from the Cambridge Polymer Group in Cambridge, MA. Data analysis is discussed by Szabo [69].

Results for the growth of tensile stress during the constant-stretch-rate extension of a constant-viscosity, ideal elastic liquid at a variety of stretch rates are shown in Figure 7.14 [68]. The similarity of these results to the LDPE melt data presented in Figure 7.1 is unmistakable. In general, data for the different stretch-rate runs superpose at short times; during this time, the ratio of the stress growth coefficient to the shear viscosity attains a value of 3 and remains so for varying periods, depending on the extension rate imposed. This is then followed by a dramatic increase in the extensional stress as the fluid is stretched further; the deviation from linear viscoelastic behavior and commencement of strain hardening is noticeable at total strain values of about unity. The total strain in these experiments is often as large as 7, a value comparable to that achieved in the melt stretching runs. Steady states are clearly achieved at high stretch rates, and the Trouton ratio approaches 10,000. Thus, the measured Trouton ratios for this polymer solution are easily about two orders of magnitude larger than those for the most extension-thickening polymer melts. Similar results are obtained on other constant-viscosity liquids, such as M1, and also on the shear-thinning solution A1 [68, 70].

Regarding the use of rheological constitutive equations to describe the observed stress growth behavior of polymer solutions, we must remember to include the contribution to the stress made by the solvent. Although this contribution may not always be significant in extensional flow, it can have a major influence on shear behavior; if the solvent contribution is not included, errors will arise in the parameter values if these are obtained by fitting shear data [71]. As a consequence, the simplest constitutive equation for describing polymer solution viscoelasticity is a

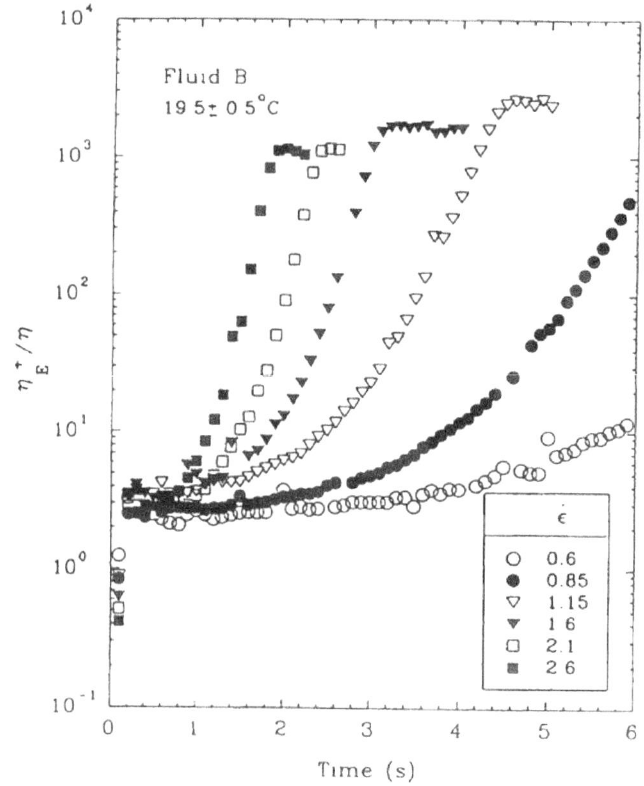

FIGURE 7.14 Tensile stress growth data for the constant-viscosity, elastic liquid of Figure 7.12. (From Ref. [68].)

sum of Eq. (7.3) written for the polymer contribution and the Newtonian equation written for the solvent contribution; the resulting three-constant equation is commonly called the *Oldroyd model B* [71]. Tirtaatmadja and Sridhar [70] found that the Oldroyd B equation written using a four-element discrete relaxation time spectrum could describe the shear rheology of the constant-viscosity elastic liquids as well as the growth of tensile stress in uniaxial extension. However, owing to the inherent limitations of the Maxwell equation, this model cannot predict a steady state in the extensional stress. For the shear-thinning polymer solution Al, the use of the Oldroyd B model is clearly inappropriate, and here the White-Metzner equation was found to do a good job overall. This equation is the same as Eq. (7.3), but the two parameters are not constants; these are functions of the invariants of the rate-of-deformation tensor [72].

Later work was done by Sridhar and his students [73, 74] on solutions of several nearly monodisperse, high-molecular-weight polystyrenes dissolved in a theta-like solvent made up of a mixture of low-molecular-weight polystyrene and dioctyl phthalate. In essentially each case, it was found that all the curves of the type shown in Figure 7.14 could be reduced to a single curve if data were plotted as a function of strain. There was an initial linear region that gave way to a plateau at a strain of 0.2–0.4 units; here, the transient Trouton ratio had a value of 3. Beyond this, strain hardening occurred at a strain of 2–3, and a steady state was attained when the strain reached about 5.5. In some cases, the steady state value decreased as the strain rate increased. This is reminiscent of the behavior of LDPE melts shown in Figure 7.7, and the slope is again −0.5 [74, 75]. As far as the effect of polymer concentration is concerned, the steady extensional viscosity is found to be proportional to the concentration, and this is similar to the behavior of the steady shear viscosity of dilute polymer solutions. The effect of changing molecular weight was not as clear-cut, but it appeared that, in most cases, extensional data scaled with molecular weight as well. Reference [75] may be consulted for a review of filament stretching rheometry.

Related to the filament stretching rheometer is the *filament breakup rheometer* [76]. Here, a step strain is imposed on the fluid sample that is enclosed between the two disks, and the inter-disk distance is held constant. As time increases, the filament thins, and the mid-filament diameter is monitored; the kinematics are found to be those of extensional flow, and Hencky strains as large as 12 can be achieved. The stretch rate can be computed from a measurement of the filament diameter, and it is found to depend on time. The transient extensional viscosity can be calculated if one is able to estimate the axial force in the thinning filament. This is done with the help of the Laplace equation as the ratio of the coefficient of surface tension divided by the instantaneous filament radius [77]. It is found that the extensional viscosity calculated in this manner asymptotically approaches the steady state extensional viscosity determined by the filament stretching technique [76]. An instrument, based on the filament thinning principle and christened Capillary breakup extensional rheometer (CaBER), was commercialized by the Cambridge Polymer Group in 2002. For very low viscosity samples, the time scale of the experiment can become comparable to the time needed to impose the step strain, and this can introduce errors. One way to avoid this is to use the *dripping-onto-substrate* (DoS) technique, where a liquid bridge is formed by the slow dripping of a single liquid drop onto a substrate [78].

III. EXTENSIONAL VISCOSITY AT HIGH STRETCH RATES

The major limitation of the uniaxial extensional rheometers described so far is their inability to make measurements on a greater variety of fluids and at stretch rates characteristic of polymer processing operations. Other techniques are therefore needed to stretch polymeric liquids at stretch rates of the order of 100 sec^{-1}. Invariably, these methods involve nonuniform stretching—neither the stress nor the stretch rate is constant in the Lagrangian sense. In addition, the fluid is rarely in a virgin, stress-free state to begin with. In this regard, fiber spinning and converging flow have proven to be especially popular. However, from stresses measured in these flow fields, we cannot obtain true extensional viscosity as defined by the equations given in Chapter 1. Nonetheless, data from properly designed nonuniform stretching experiments can, in principle, be analyzed with the help of rheological constitutive equations, i.e., results can be employed to obtain the constants in a constitutive equation, which can, in turn, be used to predict the extensional viscosity. In addition, we can compute the ratio of instantaneous tensile stress to instantaneous stretch rate. This ratio is termed an *apparent extensional viscosity,* and it provides a simple measure of the resistance that polymeric fluids offer to extensional deformation.

A. Fiber Spinning

Polymer *melt spinning* is one of the major processes for the manufacture of synthetic fibers, and different aspects of this operation may be found in Chapter 15 of Ref. [55]. A simplified fiber spinning setup is shown in Figure 7.15, which displays a schematic of the commercially available Rheotens tensile tester for polymer melts, made by Göttfert in Germany [30]. A capillary viscometer or a laboratory extruder pumps molten polymer through a die of circular cross-section, and the melt strand is pulled vertically downward at a constant velocity by two counterrotating, knurled wheels. For isothermal testing, the fiber may be enclosed in a constant-temperature oven, and the drawdown force is typically measured using a tensiometer. In the Rheotens instrument, the pulling speed can also be increased from rest at a constant rate until the filament breaks; the maximum velocity is 1.2 m/sec, and the maximum acceleration is 12 cm/sec^2. Representative results on the stretching of an LDPE sample in room temperature air are shown in Figure 7.15 [30]. The speed at fiber rupture is indicative of melt extensibility, and the fracture force is judged to be the *melt strength*.

For the purpose of data analysis in a constant-pulling-speed experiment, one assumes that the liquid is incompressible, the flow is steady and axisymmetric, and the velocity is uniform over the cross-section. Under these conditions, the fluid

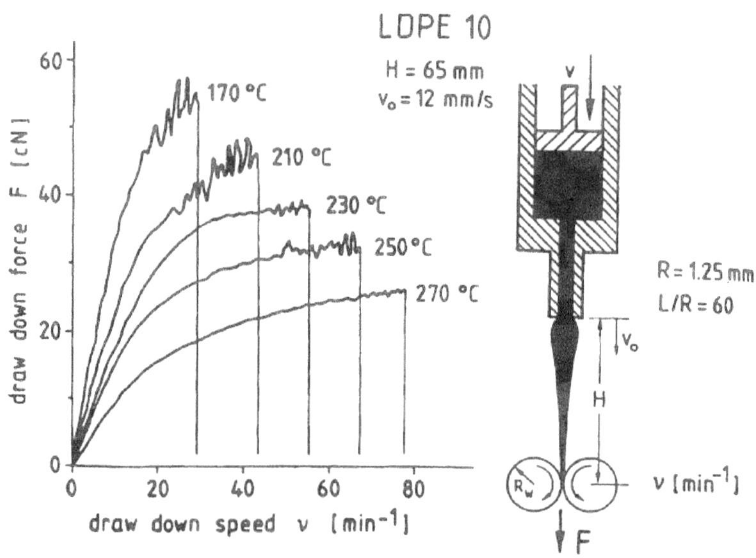

FIGURE 7.15 Representative tensile testing results obtained using the Rheotens instrument shown on the right of the figure. (From Ref. [30].)

velocity depends only on position down the spinline, and the velocity as well as the velocity gradient can be computed from a knowledge of the volumetric flow rate and the experimentally determined diameter profile. The viscoelastic fluid stress at any location is then calculated from the measured pulling force through a momentum balance [1, 3, chapter 15 of 55]. The ratio of the stress to the stretch rate at the same location is plotted either against the stretch rate or against position down the spinline. An example of such a graph is given in Figure 7.16 for the isothermal spinning of a low-density polyethylene at three different temperatures [79]; the apparent elongational viscosity is plotted versus the stretch rate, which, in this case, increases with position in the flow direction. It is evident that (1) the stretch rates achieved are not higher than those in the filament stretching runs of Laun and Münstedt [7, 35], and (2) although the Trouton ratios are similar to those shown in Figures 7.7 and 7.8, the shape of the extensional viscosity curves is quite different. Concerning these observations, one generally has to go to nonisothermal spinning in order to attain high stretch rates, and the kind of data shown in Figure 7.16 cannot be expected to agree quantitatively with rod-pulling results, because steady states are rarely, if ever, achieved in fiber spinning. It is only for Newtonian liquids that the apparent extensional viscosity equals the true extensional viscosity. Still, fiber spinning data are useful, and these may be utilized to determine the constants in any candidate constitutive equation [80]; the equation can then be used for process analysis and simulation.

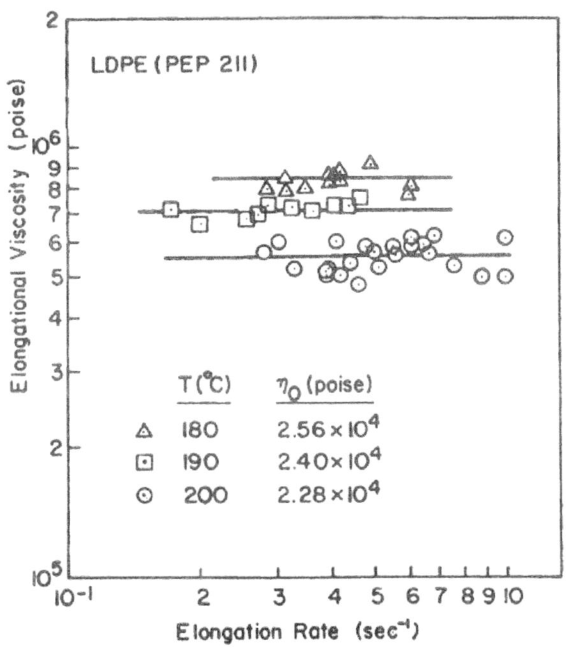

FIGURE 7.16 Apparent elongational viscosity versus elongation rate for the fiber spinning of an LDPE sample at three different temperatures. (From Ref. [79].)

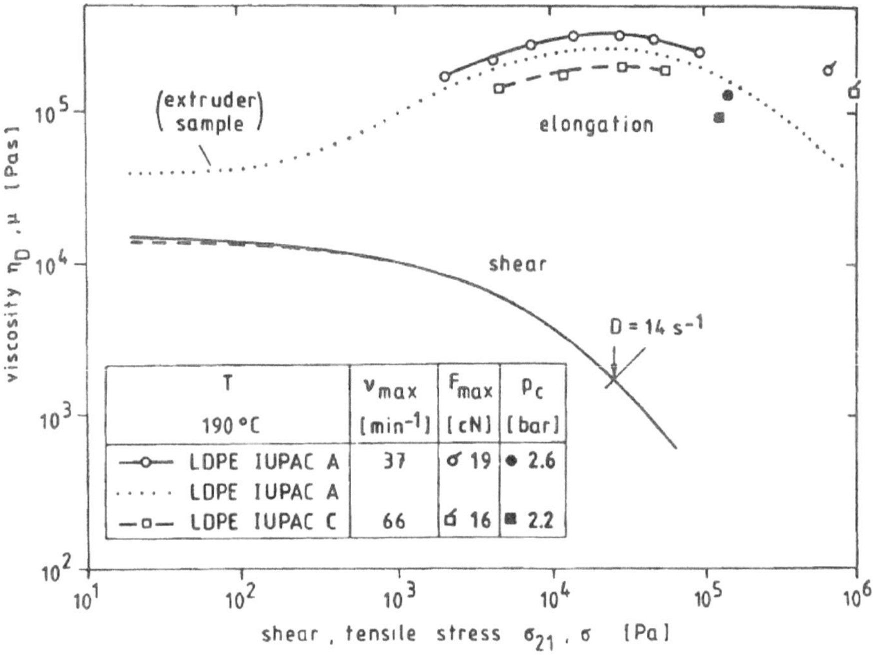

FIGURE 7.17 Comparison of different methods of extensional viscosity measurement. Unfilled symbols with a tick mark are Rheotens data; filled symbols are the results of converging-flow experiments. (From Ref. [30].)

On going to nonisothermal fiber spinning, we find that the measured force is essentially the viscoelastic force in the fiber. Recognizing this fact, Laun and Schuch [30] simply divided the measured force by the fiber cross-sectional area at the end of the spinline to obtain the extensional stress there. They then computed the apparent (nonisothermal) extensional viscosity at that location and compared it with the true extensional viscosity measured at the temperature at which the polymer leaves the barrel of the capillary viscometer. Results on three IUPAC LDPE samples are shown in Figure 7.17 [30]. These three materials are from the same batch, and their shear behavior is essentially identical. Differences in sample preparation, however, result in differences in extensional behavior. Not surprisingly, the nonisothermal Rheotens data do not superpose with the isothermal, homogeneous filament stretching data, but they appear to reveal the right trends. Also, the order of magnitude of the apparent extensional viscosity is correct. This simple test, therefore, seems to have practical utility. The constitutive modeling of Rheotens data may be found in Ref. [81].

In the fiber spinning of polymer solutions, as in the case of polymer melts, a fluid jet emerging from a nozzle is stretched, and the resulting tensile force is measured. A variety of devices have been proposed over the years [3]; Figure 7.18 shows the spinline viscometer of Ferguson and Hudson [82], which was the basis of a commercial rheometer manufactured by Carri-Med (now TA Instruments) in the United Kingdom. In this instrument, fluid is stretched when it comes into contact with a

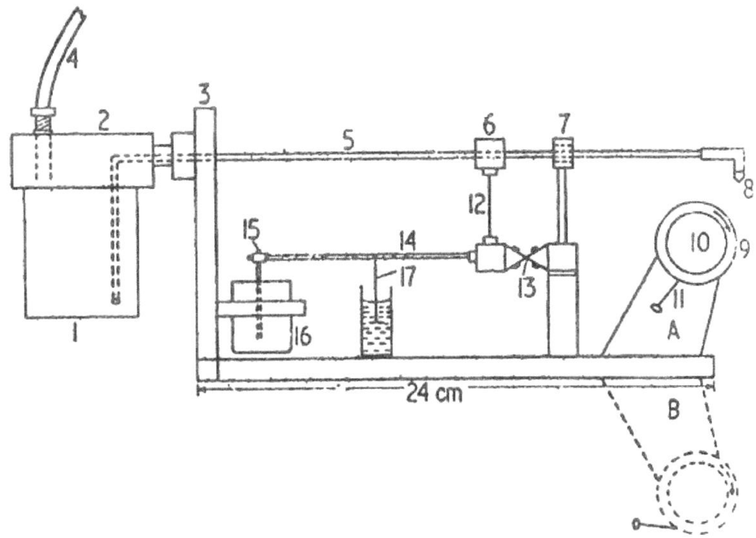

FIGURE 7.18 The spinline viscometer. 1, fluid reservoir; 5, metal tube; 8, nozzle assembly; 9, rotating drum; 11, scraper plate; 13, pivot assembly; 16, inductive displacement transducer; 17, damper. (From Ref. [82]. Used by permission of Institute of Physics Publishing, Bristol, U.K.)

rotating drum, and a variation in stretch rates is achieved by varying the speed of rotation of the drum. With increasing speed of rotation, though, there is slip between the drum and the liquid stream, and this places an upper limit on the achievable stretch rate. The force (up to 1 g) generated by stretching of the polymer solution is measured by the deflection of a thin-walled tube through which the fluid passes. The entire assembly is housed inside a heated environmental chamber, and this allows for operation to a temperature of 100°C. Data analysis for solutions is identical to that for melts except that the contributions to the measured force from fluid inertia, surface tension, gravity, and air drag can, in general, not be neglected.

A version of the spinline rheometer that can be used to make extensional flow measurements on extremely low-viscosity polymer solutions has been developed by Sridhar and Gupta, and this is shown in Figure 7.19 [83]. In this instrument, liquid enters a tank at a constant flow rate, compresses air to some steady-state value, and leaves through a capillary. When the liquid strand exiting the capillary is stretched, the air pressure in the tank drops to a new steady-state value, and this is measured using a pressure transducer. A simple force balance shows that the decrease in pressure equals the tensile stress in the liquid at the capillary outlet [83]. The tensile stress at other axial locations along the fiber is again obtained via a momentum balance. Since this viscometer has no moving parts, and since pressure measurements can be made with great ease and accuracy, apparent extensional viscosities of even very dilute aqueous polymer solutions can be determined [84]. Note that in this instrument, fluid stretching is accomplished by using a suction device in which the liquid emerging from the capillary is sucked into another capillary of still smaller

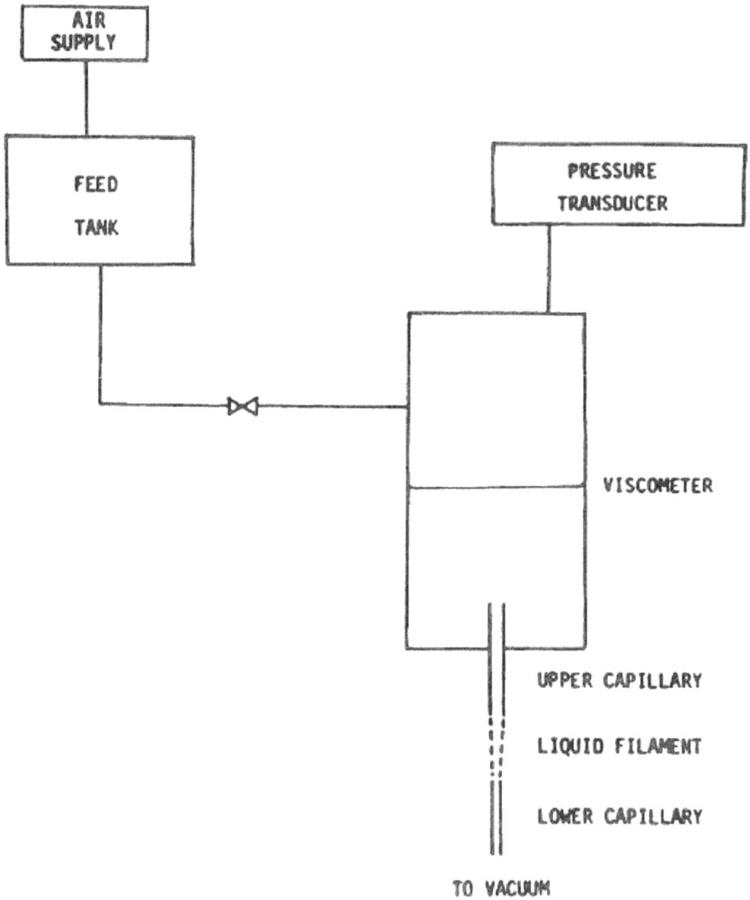

FIGURE 7.19 Schematic diagram of the extensional viscometer developed by Sridhar and Gupta. (From Ref. [83]. Used by permission of Steinkopff Publishers, Darmstadt, FRG.)

diameter, producing a stretched jet in the air gap between the capillaries [85]; stretch rates in excess of 1000 sec^{-1} can be attained [84]. At these high stretch rates, however, it is very easy to cause polymer chain scission [86], and it is advisable not to recycle the solution in the process of extensional viscosity measurement.

A different version of the instrument shown in Figure 7.19 was developed by Agarwal and Gupta [87, 88], and this is displayed in Figure 7.20 [89]. The operating principle of this device is based on the recognition that if the viscometer in Figure 7.19 is open to the atmosphere, stretching of the liquid filament again results in a tensile stress at the exit of the upper capillary, and, if the liquid level is maintained constant, this is equivalent to an increase in the pressure drop across the upper capillary. As a consequence, the flow rate through the upper capillary increases. The flow rate, with and without fluid stretching, is determined by

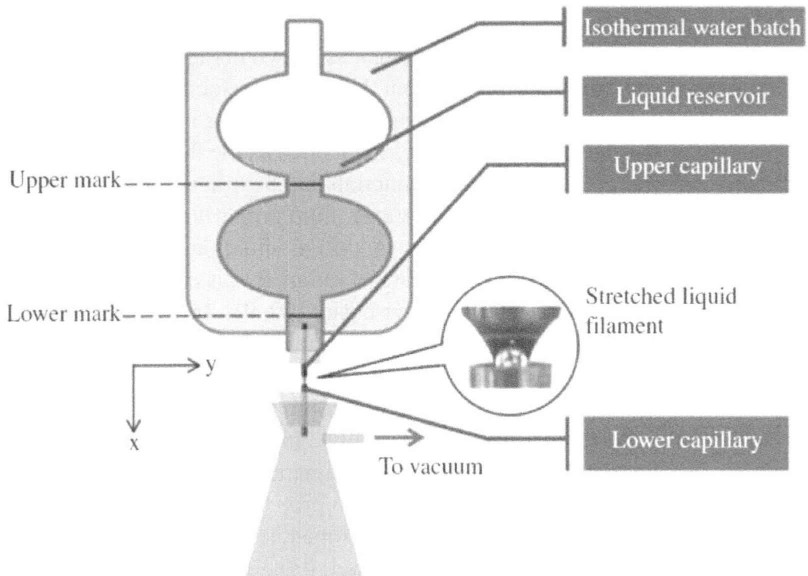

FIGURE 7.20 Schematic diagram of the extensional viscometer developed by Gupta and Agarwal. (From Ref. [89].)

measuring drainage times using a procedure similar to that employed for the measurement of intrinsic viscosity using a glass capillary viscometer [55]. Since there is a one-to-one relationship between the volumetric flow rate and the pressure drop for the Poiseuille flow of a power-law liquid through a tube of circular cross-section, a change in flow rate translates to a change in pressure drop; this pressure change equals the tensile stress at the exit of the upper capillary. The tensile stress at any other point along the stretched liquid filament is computed with the help of a force balance as before, while the filament shape allows for the calculation of the stretch rate. Clearly, only a small amount of liquid is required for the measurement of the apparent extensional viscosity.

Han et al. [89] made extensional flow measurements on a series of low-molecular-weight polystyrenes dissolved in DECALIN (cis- and trans-decahydronaphthalene). The polymer concentration was adjusted to give a solution viscosity of only 5 mPa-s, and fluid elasticity levels were so low that they could not be detected through storage modulus measurements. The resulting Trouton ratios were correlated with the jetting behavior of these solutions using drop-on-demand jetting equipment [89]. It was found that the use of Newtonian liquids (no elasticity) resulted in the formation of undesirable satellite drops. On the other hand, solutions having a Trouton's ratio in excess of 5 did not separate cleanly from the nozzle and exhibited long tails. The use of this viscometer demonstrated that the optimum range for stable and efficient inkjet printing occurred at Trouton's ratios in a narrow band between 3 and 5.

B. Converging Flow

When a liquid flows through a conical channel, as shown in Figure 7.21, there is obviously shearing of the fluid near the wall, but the average velocity increases monotonically as fluid elements move toward the apex. This increase in velocity is the result of decreasing cross-sectional area for flow, and it endows the flow field with a significant extensional character. The relative amounts of shear and extension depend on the channel geometry, the volumetric flow rate, fluid properties, and the presence of any low-viscosity lubricant near the walls. A similar situation arises in the entrance region leading from a reservoir to a capillary of radius R; this can be exploited for the measurement of the apparent extensional viscosity [90, 91]. In yet another variation, Padmanabhan and Bhattacharya have made extensional flow measurements on fluids flowing through a rectangular cross-section abrupt contraction [92]. This method gives the planar extensional viscosity rather than the uniaxial extensional viscosity. Planar extension is the same as pure shear, defined in Chapter 1.

What we measure in this flow field are the volumetric flow rate Q and the entrance pressure drop, P_{ent}, defined as the total pressure drop across the abrupt contraction between two points where the flow is fully developed, less the pressure drop for fully developed Poiseuille flow. According to Cogswell [90, 91], P_{ent} can be separated into a shear component and an extensional component. The latter contribution can be used to compute the average net tensile stress in the entry region. The result is:

$$\sigma_L = \left(\frac{3}{8}\right)(n+1)P_{\text{ent}} \tag{7.6}$$

in which n is the power-law index for shear flow. The average stretch rate is calculated to be [93]:

$$\dot{\varepsilon} = \frac{4\tau\dot{\gamma}}{3(n+1)P_{\text{ent}}} \tag{7.7}$$

where τ is the shear stress at a shear rate $\dot{\gamma}$ equal to $4Q/(\pi R^3)$. Dividing the average net tensile stress by the average stretch rate gives an average extensional viscosity.

A much more rigorous analysis of obtaining the extensional viscosity from *converging flow* measurements has been provided by Binding [94], but the method of

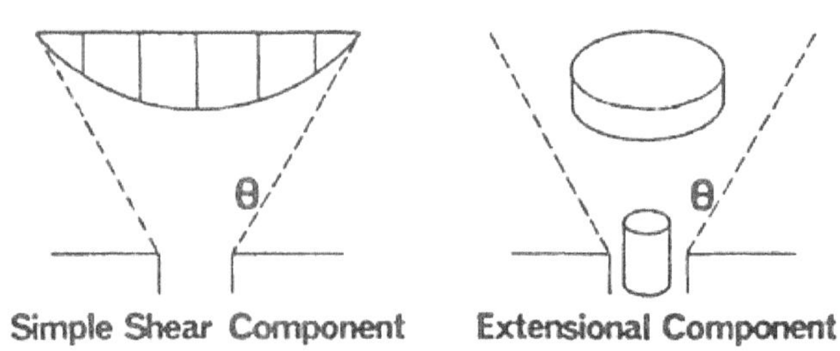

FIGURE 7.21 Converging flow into a conical channel. (From Ref. [90].)

Cogswell is simpler to use and, perhaps, just as accurate. Equations for the calculation of the uniaxial extension viscosity and the planar extension viscosity using both methods may be found in [92, 95].

Laun and Schuch [30] used the converging-flow method to measure the extensional-flow properties of two IUPAC LDPE samples, and the results are shown in Figure 7.17. It appears that the technique not only ranks the polymer melts correctly but also gives data that agree with the true extensional viscosities. Indeed, Kwag and Vlachopoulos [96] used the data of Laun and Schuch in a finite element simulation and concluded that Cogswell's method was an acceptable approximation at high stretch rates. Although this method poses no experimental problems for work with either polymer melts or low-viscosity polymer solutions, it should be kept in mind that increasing the stretch rate by increasing the volumetric flow rate results in a reduction in the fluid residence time in the converging capillary; as a consequence and keeping in view the data of Figure 7.1, the measured average-extensional-viscosity is likely to become progressively smaller than the true extensional viscosity with increasing stretch rate [97]. For the purpose of comparing the extensional viscosity obtained from entrance pressure drop measurements with the corresponding value determined from the fiber-windup technique, therefore, it has to be ensured that the value of the total strain is the same in the two experiments. This was done by Padmanabhan and Macosko using different LDPE samples, and results for IUPAC-X are shown in Figure 7.22 [95].

Here, entrance pressure drop measurements were made with a capillary rheometer fitted with a circular cross-section die that gave a contraction ratio of 12:1. The

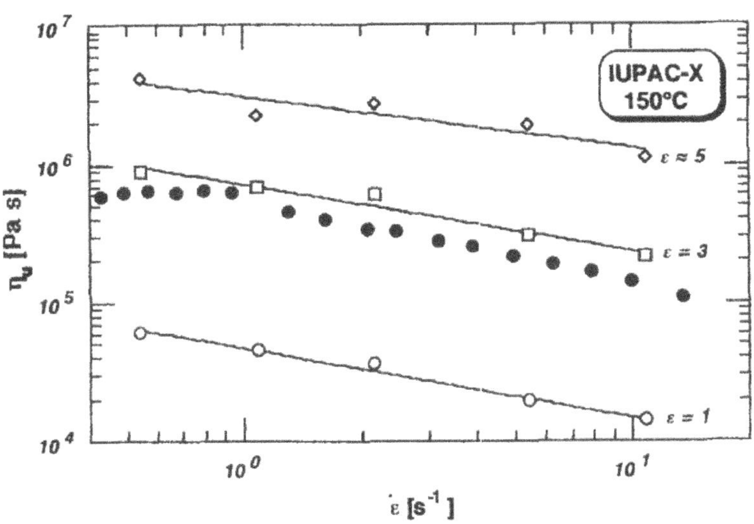

FIGURE 7.22 Comparison of uniaxial extensional viscosities from the entrance pressure drop method (filled symbols) and the transient extensional viscosities (open symbols) from the fiber wind-up method at different values of the strain for IUPAC-X LDPE. The lines shown are power-law fits to the fiber-windup data. (From Ref. [95].)

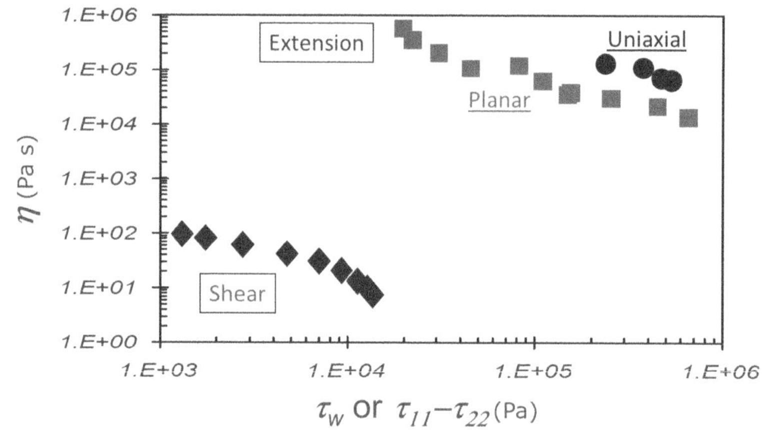

FIGURE 7.23 Shear and extensional viscosities for a hard wheat dough containing 70% added water. (Figure adapted from Ref. [4].)

fiber-windup data are on the same polymer, and these were published earlier [28]. It is seen that there is very good agreement between the two sets of results at a total strain of about 3. This agreement offers confidence that results obtained with the use of the entrance pressure drop method will be reliable even at higher deformation rates, where comparison with the fiber-windup data is not possible.

Commercial capillary rheometers have small-diameter barrels. When loading highly viscoelastic materials, such as wheat dough, into the barrel, it is very easy to trap large quantities of air. To avoid this, one can use a capillary rheometer with a large-diameter barrel. With such an apparatus, Padmanabhan [4] measured uniaxial and planar extensional viscosities for a wheat dough, by extruding the dough through dies of circular and rectangular cross-section, with abrupt contractions. The results, shown in Figure 7.23, demonstrate that the uniaxial extensional viscosity is higher than the planar extensional viscosity, due to uniaxial extensional flow being a stronger flow [6]. Results for two different doughs of different water contents are shown in Figure 7.24. Padmanabhan and Bhattacharya [98] reported planar extensional viscosity of corn meal dough during extrusion using rectangular cross-section abrupt contraction.

As mentioned earlier, Binding [99] has carried out a very sophisticated analysis of contraction flows, and commercial capillary instruments such as the Rosand RH7 twin-bore viscometer (Rosand Precision, United Kingdom) are now available for making measurements as a function of both stretch rate and temperature. This technique has also been used to evaluate the influence of pressure on the elongational properties of polymer melts [100]. A variation of the method described above is a converging hyperbolic slot die positioned in-line with an extruder. The hyperbolic shape gives rise to a constant average stretch rate, and this is the principle of the Leistritz elongation rheometer. Tensile stresses are determined using readings from standard pressure sensors. The residence time in the measuring device ranges between one and three minutes, and stretch rates as high as 75 s^{-1} can be attained.

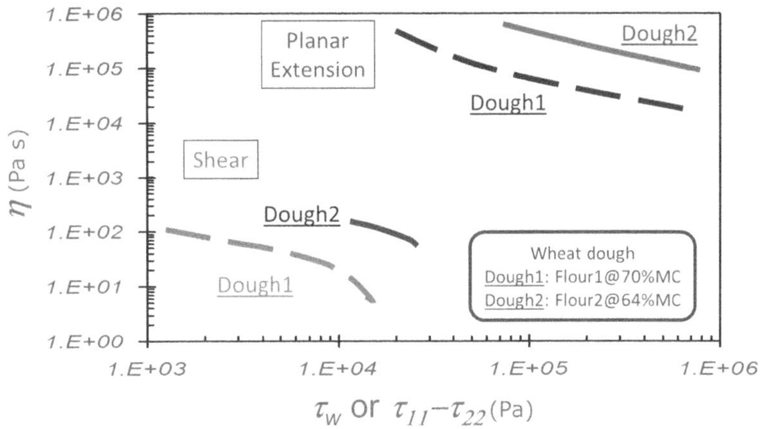

FIGURE 7.24 Shear and extensional viscosities for two wheat doughs. (Figure adapted from Ref. [4].)

C. Opposed-Nozzle Rheometer

Fuller et al. [101] have described an extensional viscometer for mobile liquids that uses two *opposed nozzles* attached to tubes immersed in a beaker of the test fluid. As shown in Figure 7.25, one of the tubes is fixed, and the other is mounted on a knife edge and is free to swing under the influence of a torque that arises from the fluid forces between the nozzles. The force of interest is a tensile force generated when fluid is sucked into the nozzles by the application of vacuum, creating an approximate extensional flow, with a stretch rate as large as 10,000 sec^{-1}, in the small gap between the nozzles. This instrument was manufactured by Rheometrics Scientific under the name Fluids Extensional Analyzer, and it provides an average extensional viscosity; the tensile stress is calculated via the measured torque, and the ratio of the average fluid velocity in the nozzle to half the nozzle separation is taken to be the stretch rate. It is often found that measurements on Newtonian liquids give a value of the extensional viscosity that is slightly different from three times the shear viscosity. The flow field between the opposed nozzles has been thoroughly analyzed by Scriven and coworkers [102, 103], who have concluded that "opposed nozzles should be considered as useful indexers rather than rheometers". Padmanabhan has reviewed the fiber windup method, the entrance pressure drop method, and the opposed jets device and compared the pros and cons of these different techniques for the measurement of extensional viscosity of viscoelastic liquid foods; each method is relevant for specific foods in specific ranges of process conditions [4].

Other stagnation point devices within which one may create well-defined extensional flow are cross-slot devices, where one can get either uniaxial or biaxial extensional flow [104]. The use of micro-particle image velocimetry yields the stretch rate, while excess pressure drop measurements give the needed tensile stresses [105]. Data have been obtained on dilute polymer solutions.

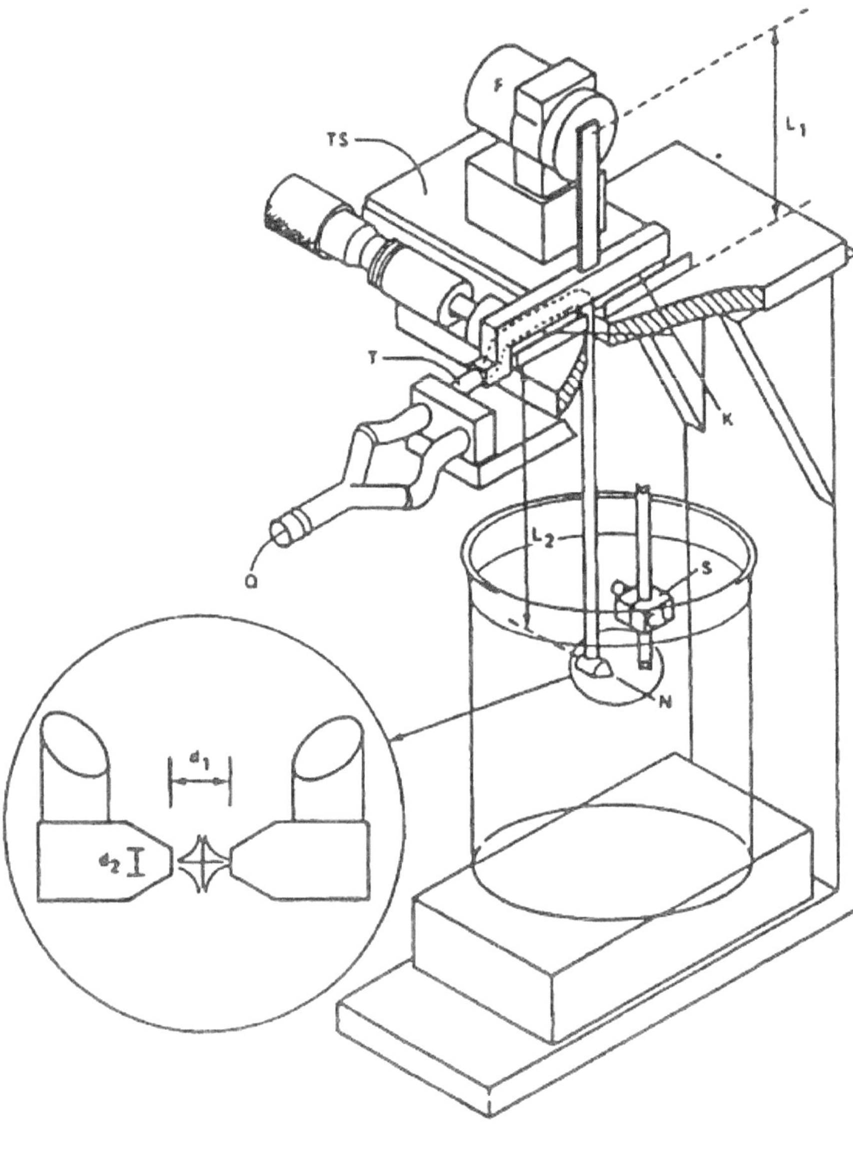

FIGURE 7.25 Opposed-nozzle rheometer. N, opposing nozzles; S, fixture defining nozzle separation; K, knife edge fulcrum; F, force transducer; L, L_2, lever arms. (From Ref. [101].)

IV. CONCLUDING REMARKS

It is now well established that when the mode of deformation in a polymer processing situation is predominantly extensional in character, it is the extensional viscosity of the polymer, rather than the shear viscosity, that is the better predictor of observed behavior. This has been established, for example, in film blowing [106] and blow molding [107], even when the extensional viscosity measurements were made at relatively low stretch rates. These measurements are well understood in terms of their dependence on molecular parameters such as molecular weight, molecular weight distribution, and long-chain branching; data on well-characterized melts and solutions have also been used to test the validity of different proposed rheological constitutive equations. However, there is still a pressing need to develop techniques capable of measuring the true extensional viscosity of polymeric liquids at deformation rates more representative of industrial polymer processing. Extensional rheology is also important for the structural characterization of liquid foods, especially as they contain proteins and polysaccharides that are polymeric. It should also be recognized that the sensory perception of food in the mouth includes an extensional component during, for example, sucking of a beverage through a straw, mouth-feel, and swallowing of beverages [4].

Even though in this chapter we have mainly described only those measurement techniques that have led to commercial instruments, many ingenious approaches have been tried in the past [1–3, 108–110]. We mention that, in addition to the uniaxial extension viscosity, we also measure an equal biaxial extension viscosity and a planar extension viscosity [6, 108–110]. Popular methods for experimentally determining biaxial extension viscosity include bubble inflation [6, 111, Appendix], lubricated squeezing [6, 59, 112], and sheet stretching [113, 114]. These are considered in detail in the chapter on the flow behavior of suspensions.

REFERENCES

1. C. J. S. Petrie. Elongational Flows. Pitman, London, 1979.
2. M. Rides, R. G. Allen, and S. Chakravorty. Review of Extensional Viscoelasticity Measurement Techniques for Polymer Melts, Report #CMMT(A)44, National Physical Laboratory, Teddington, 1996.
3. R. K. Gupta, and T. Sridhar. Elongational rheometers. In: A. A. Collyer, D. W. Clegg, eds. Rheological Measurement. 2nd ed., Chapman and Hall, London, 1998, pp. 516–549.
4. M. Padmanabhan. Measurement of extensional viscosity of viscoelastic liquid foods. J. Food Eng. 25:311–327 (1995).
5. R. J. Good and R. K. Gupta. The coupling of interfacial, rheological, and thermal control mechanisms in polymer adhesion. In: L.-H. Lee, ed. Adhesive Bonding. Plenum Press, New York, 1991, pp. 47–73.
6. M. Padmanabhan. Rheology: Extensional. In: D. R. Heldman, C. I. Moraru, eds. Encyclopedia of Agricultural, Food, and Biological Engineering. 2nd ed. Taylor & Francis, Boca Raton, FL, 2011, pp. 1473–1479.
7. H. Münstedt and H. M. Laun. Elongational behavior of a low density polyethylene melt. II. Transient behavior in constant stretching rate and tensile creep experiments. Comparison with shear data. Temperature dependence of the elongational properties. Rheol. Acta 18:492–504 (1979).
8. R. L. Ballman. Extensional flow of polystyrene melt. Rheol. Acta 4:137–140 (1965).
9. F. N. Cogswell. Tensile deformations in molten polymers. Rheol. Acta 8:187–194 (1969).
10. L. Nicodemo, B. De Cindio, and L. Nicolais. Elongational viscosity of microbead suspensions. Polym. Eng. Sci. 15:679–683 (1975).
11. J. Vlachopoulos and N. Polychronopoulos. Basic concepts in polymer melt rheology and their importance in processing. In: M. Kontopoulou, ed. Applied Polymer Rheology: Polymeric Fluids with Industrial Applications. Wiley, Hoboken, 2012, pp. 1–27.
12. L. E. Nielsen. Polymer Rheology. Marcel Dekker, New York, 1977.

13. P. Micic, S. N. Bhattacharya, and G. Field. Rheological behavior of linear low density polyethylene/low density polyethylene blends under elongational deformation—The effect of molecular structure and melt morphology. Intern. Polym. Processing 12:110–115 (1997).
14. S. J. Armour, J. C. Muirhead, and A. B. Metzner. Filament formation in viscoelastic fluids. Proc. X Intl. Congress on Rheology, Acapulco, Mexico, pp. 143–151, 2/1984.
15. J. E. Matta and R. P. Tytus. Viscoelastic breakup in a high velocity airstream. J. Appl. Polym. Sci. 27:397–405 (1982).
16. J. E. Glass. Dynamics of roll spatter and tracking, Part III. Importance of extensional viscosities. J. Coatings Technol. 50(641):56–71 (1978).
17. D. F. Massouda. Analysis and prediction of roll spatter from latex paints. J. Coatings Technol. 57(722):27–36 (1985).
18. D. A. Soules, R. H. Fernando, and J. E. Glass. Dynamic uniaxial extensional viscosity (DUEV) effects in roll application I. Rib and web growth in commercial coatings. J. Rheol. 32:181–198 (1988).
19. E. K. Hadde, J. A. Y. Cichero, S. Zhao, W. Chen, and J. Chen. The importance of extensional rheology in bolus control during swallowing. Sci. Rep. 9:16106 (2019).
20. J. Rhi-Sausi and J. Dealy. An extensiometer for molten plastics. Polym. Eng. Sci. 16: 799–802 (1976).
21. P. K. Agrawal, W. K. Lee, J. M. Lorntson, C. I. Richardson, K. F. Wissbrun, and A. B. Metzner. Rheological behavior of molten polymers in shearing and in extensional flows. Trans. Soc. Rheol. 21:355–379 (1977).
22. F. N. Cogswell. The rheology of polymer melts under tension. Plast. Polym. 36:109–111 (1968).
23. G. V. Vinogradov, V. D. Fikhman, and B. V. Radushkevich. Uniaxial extension of polystyrene at true constant stress. Rheol. Acta 11:286–291 (1972).
24. H. Münstedt. Viscoelasticity of polystyrene melts in tensile creep experiments. Rheol. Acta 14:1077–1088 (1975).
25. J. Meissner. Rheometer zur Untersuchung der deformationsmechanischen Eigenschaften von Kunststoff-Schmelzen unter definierter Zugbeanspruchung. Rheol. Acta 8:78–88 (1969).
26. J. Meissner and J. Hostettler. A new elongational rheometer for polymer melts and other highly viscoelastic liquids. Rheol. Acta 33:1–21 (1994).
27. D. W. de Bruijne, J. de Loof, and A. van Eulem. The rheological properties of bread dough and their relation to baking. In: R. E. Carter, ed. Rheology of Food, Pharmaceutical and Biological Materials with General Rheology. Elsevier, London, 1990, 269–283.
28. M. Padmanabhan, L. Kasehagen, and C. Macosko. Transient extensional viscosity from a rotational shear viscometer using fiber-windup technique. J. Rheol. 40:473–481 (1996).
29. C. W. Macosko and J. M. Lorntson. The rheology of two blow molding polyethylenes. SPE Tech. Papers 19:461–467 (1973).
30. H. M. Laun and H. Schuch. Transient elongational viscosities and drawability of polymer melts. J. Rheol. 33:119–175 (1989)
31. M. L. Sentmanat. Miniature universal testing platform: From extensional melt rheology to solid-state deformation behavior. Rheol. Acta 43:657–669 (2004).
32. M. Sentmanat, B. N. Wang, and G. H. McKinley. Measuring the transient extensional rheology of polyethylene melts using the SER universal testing platform. J. Rheol. 49:585–606 (2005).
33. T. S. Ng, G. H. McKinley, and M. Padmanabhan. Linear to non-linear rheology of wheat flour dough. Appl. Rheol. 16:265–274 (2006).
34. J. Meissner. Basic parameters, melt rheology, processing and end-use properties of three similar low density polyethylene samples. Pure Appl. Chem. 42:553–612 (1975).

35. H. M. Laun, and H. Münstedt. Elongational behavior of a low density polyethylene melt. I. Strain rate and stress dependence of viscosity and recoverable strain in the steady-state. Comparison with shear data. Influence of interfacial tension. Rheol. Acta 17:415–425 (1978).
36. T. Raible, A. Demarmels, and J. Meissner. Stress and recovery maxima in LDPE melt elongation. Polymer Bull. 1:397–402 (1979).
37. R. Zullo and S. Iannace. The effects of different starch sources and plasticizers on film blowing of thermoplastic starch: Correlation among process, elongational properties and macromolecular structure. Carbohydr. Polym. 77:376–383 (2009).
38. H. Münstedt. Dependence of the elongational behavior of polystyrene melts on molecular weight and molecular weight distribution. J. Rheol. 24:847–867 (1980).
39. A. Franck and J. Meissner. The influence of blending polystyrenes of narrow molecular weight distribution on melt creep flow and creep recovery in elongation. Rheol. Acta 23:117–123 (1984).
40. J. Linster and J. Meissner. Melt elongation and structure of linear polyethylene (HDPE). Polymer Bull. 16:187–194(1986).
41. V. S. Au-Yeung, C. W. Macosko, and V. R. Raju. Extensional flow of linear and star-branched hydrogenated polybutadiene with narrow molecular weight distribution. J. Rheol. 25:445–452 (1981).
42. R. A. Bubeck. Structure-property relationships in metallocene polyethylenes. Mater. Sci. Eng. R. 39:1–28 (2002).
43. A. Williamson, J. Gomes, T. Karjala, and C. do Valle. Rheology modifier additive for enhanced processability of polyethylene for blown-film applications. J. Vinyl Addit. Technol. 29:703–723 (2023).
44. H. Münstedt and H. M. Laun. Elongational properties and molecular structure of polyethylene melts. Rheol. Acta 20:211–221 (1981).
45. F. P. La Manda, A. Valenza, and D. Acierno. Influence of long chain branching on the elongational behavior of different polyethylenes and their blends. Polymer Bull. 15:381–387 (1986).
46. H. Münstedt, S. Kurzbeck, and L. Egersdorfer. Influence of molecular structure on rheological properties of polyethylenes. Part II. Elongational behavior. Rheol. Acta 37:21–29 (1998).
47. J. K. Jorgensen, K. Redford, E. Ommundsen, and A. Stori. Molecular structure and shear rheology of long chain branched polypropylene formed by light cross-linking of a linear precursor with 1,3-benzenedisulfonyl azide. J. Appl. Polym. Sci. 106:950–960 (2007).
48. S. Cheng, E. Phillips, and L. Parks. Processability improvement of polyolefins through radiation-induced branching. Radiation Phys. Chem. 79:329–334 (2010).
49. C. Zhang, H. Niu, and J.-Y. Dong. Fabrication of long chain branched polypropylene using click chemistry. Polym. Bull. 68:949–959 (2012).
50. S. Li, M. Xiao, Y. Guan, D. Wei, H. Xiao, and A. Zheng. A novel strategy for the preparation of long chain branching polypropylene and the investigation of foamability and rheology. Eur. Polym. J. 48:362–371 (2012).
51. D. Auhl, J. Stange, H. Münstedt, B. Krause, D. Voigt, A. Lederer, U. Lappan, and K. Lunkwitz. Long-chain branched polypropylenes by electron beam irradiation and their rheological properties. Macromolecules 37:9467–9472 (2004).
52. J. Stange, C. Uhl, and H. Münstedt. Rheological behavior of blends from a linear and a long-chain branched polypropylene. J. Rheol. 49:1059–1079 (2005).
53. G. Breuer and A. Schausberger. Recovery of shear modification of polypropylene melts. Rheol. Acta 50:461–468 (2011).
54. A. S. Lodge. Constitutive equations from molecular network theories for polymer solutions. Rheol. Acta 7:379–392 (1968).

55. A. Kumar and R. K. Gupta. Fundamentals of Polymer Engineering. 3rd ed., CRC Press, Boca Raton, FL, 2019.
56. R. G. Larson. Constitutive Equations for Polymer Melts and Solutions. Butterworths, Boston, 1988.
57. M. J. Crochet, A. R. Davies, and K. Walters. Numerical Simulation of Non-Newtonian Flow. Elsevier, Amsterdam, 1984.
58. J. M. Dealy and K. F. Wissbrun. Melt Rheology and Its Role in Plastics Processing. Van Nostrand Reinhold, New York, 1990.
59. C. W. Macosko. Rheology: Principles, Measurements and Applications. VCH, New York, 1994.
60. N. Phan-Thien and R. I. Tanner. A new constitutive equation derived from network theory. J. Non-Newt. Fluid Mech. 2:353–365 (1977).
61. S. A. Khan and R. G. Larson. Comparison of simple constitutive equations for polymer melts in shear and biaxial and uniaxial extensions. J. Rheol. 31:207–234 (1987).
62. E. Narimissa and M. F. Wagner. Review on tube model based constitutive equations for polydisperse linear and long-chain branched polymer melts. J. Rheol. 63:361–375 (2019).
63. Special issue of J. Non-Newt. Fluid Mech. 30:97–368 (1988).
64. T. Sridhar. An overview of the project MI. J. Non-Newt. Fluid Mech. 35:85–92 (1990).
65. N. E. Hudson and T. E. R. Jones. The AI project—An overview. J. Non-Newt. Fluid Mech. 46:69–88 (1993).
66. Special issue of J. Non-Newt. Fluid Mech. 52:105–228 (1994).
67. T. Sridhar, V. Tirtaatmadja, D. A. Nguyen, and R. K. Gupta. Measurement of extensional viscosity of polymer solutions. J. Non-Newt. Fluid Mech. 40:271–280 (1991).
68. V. Tirtaatmadja and T. Sridhar. A filament stretching device for measurement of extensional viscosity. J. Rheol. 37:1081–1102 (1993).
69. P. Szabo. Transient filament stretching rheometer. I: Force balance analysis. Rheol. Acta 36:277–284 (1994).
70. V. Tirtaatmadja and T. Sridhar. Comparison of constitutive equations for polymer solutions in uniaxial extension. J. Rheol. 39:1133–1160 (1995).
71. G. Prilutski, R. K. Gupta, T. Sridhar, and M. E. Ryan. Model viscoelastic liquids. J. Non-Newt. Fluid Mech. 12:233–241 (1983).
72. J. L. White and A. B. Metzner. Development of constitutive equations for polymeric melts and solutions. J. Appl. Polym. Sci. 7:1867–1889 (1963).
73. R. K. Gupta. Studies on uniaxial extension of dilute polymer solutions. Ph.D. dissertation. Monash University, Clayton, Australia, 1999.
74. R. K. Gupta, D. A. Nguyen, and T. Sridhar. Extensional viscosity of dilute polystyrene solutions: Effect of concentration and molecular weight. Phys. Fluids 12:1296–1318 (2000).
75. G. H. McKinley and T. Sridhar. Filament-stretching rheometry of complex fluids. Annu. Rev. Fluid Mech. 34:375–415 (2002).
76. S. L. Anna and G. H. McKinley. Elasto-capillary thinning and breakup of model elastic liquids. J. Rheol. 45:115–138 (2001).
77. P. Schummer and K. H. Tebel. A new elongational rheometer for polymer solutions. J. Non-Newt. Fluid Mech. 12:331–347 (1983).
78. B. P. Robertson and M. A. Calabrese. Evaporation-controlled dripping-onto-substrate (DoS) extensional rheology of viscoelastic polymer solutions. Sci. Rep. 12:4697 (2022).
79. C. D. Han and R. R. Lamonte. Studies on melt spinning. Part 1. Effect of molecular structure and molecular weight distribution on elongational viscosity. Trans. Soc. Rheol. 16:447–472 (1972).
80. J. A. Spearot and A. B. Metzner. Isothermal spinning of molten polyethylenes. Trans. Soc. Rheol. 16:495–518 (1972).

81. V. Rauschenberger and H. M. Laun. A recursive model for Rheotens tests. J. Rheol. 41:719–737 (1997).
82. J. Ferguson and N. E. Hudson. A new viscometer for the measurement of apparent elongational viscosity. J. Phys. E: Sci. Instrum. 8:526–530 (1975).
83. T. Sridhar and R. K. Gupta. A simple extensional viscometer. Rheol. Acta 24:207–209 (1985).
84. R. C. Chan, R. K. Gupta, and T. Sridhar. Fiber spinning of very dilute solutions of polyacrylamide in water. J. Non-Newt. Fluid Mech. 30:267–283 (1988).
85. M. Khagram, R. K. Gupta, and T. Sridhar. Extensional flow of xanthan gum solutions. J. Rheol. 29:191–207 (1985).
86. J. A. Odell, A. Keller, and M. J. Miles. A method for studying flow-induced polymer degradation: Verification of chain halving. Polym. Commun. 24:7–10 (1983).
87. S. Agarwal and R. K. Gupta. An innovative extensional viscometer for low-viscosity and low-elasticity liquids. Rheol. Acta 41:456–460 (2002).
88. R. K. Gupta and S. Agarwal. Sensitive, spin-line type extensional viscometer for mobile liquids employing very small liquid volumes. US patent no. 7,114,374 issued 3 Oct 2006.
89. L. Han, D. Doraiswamy, and R. K. Gupta. Jetting of dilute polymer solutions. J. Therm. Anal. Calorim. 106:305–312 (2011).
90. F. N. Cogswell. Measuring the extensional rheology of polymer melts. Trans. Soc. Rheol. 16:383–403 (1972).
91. F. N. Cogswell. Converging flow of polymer melts in extrusion dies. Polym. Eng. Sci. 12:64–73 (1972).
92. M. Padmanabhan and M. Bhattacharya. In-line measurements of rheological properties of polymer melts. Rheol. Acta 33:71–87 (1994).
93. F. N. Cogswell. Converging flow and stretching flow: A compilation. J. Non-Newt. Fluid Mech. 4:23–38 (1978).
94. D. M. Binding. An approximate analysis for contraction and converging flows. J. Non-Newt. Fluid Mech. 27:173–189 (1988).
95. M. Padmanabhan and C. W. Macosko. Extensional viscosity from entrance pressure drop measurements. Rheol. Acta 36:144–151 (1997).
96. C. Kwag and J. Vlachopoulos. An assessment of Cogswell's method for measurement of extensional viscosity. Polym. Eng. Sci. 31:1015–1021 (1991).
97. R. K. Gupta and T. Sridhar. Viscoelastic effects in non-Newtonian flows through porous media. Rheol. Acta 24:148–151 (1985).
98. M. Padmanabhan and M. Bhattacharya. Planar extensional viscosity of corn meal dough. J. Food Eng. 28:389–411 (1993).
99. D. M. Binding. Contraction flows and new theories for estimating extensional viscosity. In: A. A. Collyer, ed. Techniques in Rheological Measurement. Chapman and Hall, London, 1993, pp. 1–32.
100. D. M. Binding, M. A. Couch, and K. Walters. The pressure dependence of the shear and elongational properties of polymer melts. J. Non-Newt. Fluid Mech. 79:137–155 (1998).
101. G. G. Fuller, C. A. Cathey, B. Hubbard, and B. E. Zebrowski. Extensional viscosity measurements for low-viscosity fluids. J. Rheol. 31:235–249 (1987).
102. P. R. Schunk, J. M. de Santos, and L. E. Scriven. Flow of Newtonian liquids in opposed-nozzles configuration. J. Rheol. 34:387–414 (1990).
103. P. Dontula, M. Pasquali, L. E. Scriven, and C. W. Macosko. Can extensional viscosity be measured with opposed-nozzle devices? Rheol. Acta 36:429–448 (1997).
104. S. J. Haward, F. Pimenta, S. Varchanis, D. W. Carlson, K. Toda-Peters, M. A. Alves, and A. Q. Shen. Extensional rheometry of mobile fluids. Part I: OUBER, an optimized uniaxial and biaxial extensional rheometer. J. Rheol. 67:995–1009 (2023).

105. S. J. Haward, S. Varchanis, G. H. McKinley, M. A. Alves, and A. Q. Shen. Extensional rheometry of mobile fluids. Part II: Comparison between the uniaxial, planar, and biaxial extensional rheology of dilute polymer solutions using numerically optimized stagnation point microfluidic devices. J. Rheol. 67:1011–1030 (2023).
106. A. Ghaneh-Fard, P. J. Carreau, and P. G. Lafleur. Study of kinematics of film blowing of different polyethylenes. Polym. Eng. Sci. 37:1148–1163 (1997).
107. D. H. Sebastian and J. R. Dearborn. Elongation rheology of polyolefins and its relation to processability. Polym. Eng. Sci. 23:572–575 (1983).
108. R. W. Whorlow. Rheological Techniques. 2nd ed. Ellis Horwood, New York, 1992, pp. 229–274.
109. J. M. Dealy. Extensional flow of non-Newtonian fluids—A review. Polym. Eng. Sci. 11:433–445 (1971).
110. J. M. Dealy. Extensional rheometers for molten polymers: A review. J. Non-Newt. Fluid Mech. 4:9–21 (1978).
111. D. D. Joye, G. W. Poehlein, and C. D. Denson. A bubble inflation technique for the measurement of viscoelastic properties in equal biaxial extensional flow. Trans. Soc. Rheol. 16:421–445 (1972).
112. S. Chatraei, C. W. Macosko, and H. H. Winter. Lubricated squeezing flow: A new biaxial extensional rheometer. J. Rheol. 25:433–443 (1981).
113. J. Meissner. Experimental aspects in polymer melt elongational rheometry. Chem. Eng. Commun. 33:159–180 (1985).
114. J. Meissner. Polymer melt elongation—Methods, results, and recent developments. Polym. Eng. Sci. 27:537–546 (1987).

Appendix[1]: Analysis of Alveography to Extract Biaxial Extensional Rheological Behavior

Mahesh Padmanabhan

I. INTRODUCTION

Delivering consistent product quality, during industrial manufacture of commercial products, requires consistent raw material quality. This can be achieved by using good analytical methods to measure critical properties of the raw materials and establishing specifications. Since most raw materials used in the food industry are nature-sourced, ensuring consistent raw material quality is even more challenging, placing even more stringent requirements on analytical measurements.

Wheat flour is a staple ingredient used in the manufacture of numerous baked food products such as breads, pasta, cookies, crackers, cakes, and pizza. Wheat dough rheology has long been recognized as critical to the processing and product quality of baked goods [1, 2]. This has led to the development and use of several empirical rheological tests [1, 3]. These may be broadly classified as mixing- or kneading-type instruments (e.g., Farinograph, Mixograph, and Consistograph) and extension- or stretching-type instruments (e.g., Extensograph, Alveograph, and Kiefer Extensibility rig); detailed reviews can be found elsewhere [1, 3]. Standard method protocols are available for these tests from the American Association of Cereal Chemistry (AACC), International Association for Cereal Science and Technology (ICC), and other technical organizations.

In practice, quantitative empirical parameters are typically extracted from each of the aforementioned empirical tests, in order to quantify wheat flour quality; based on years of practice and statistical empirical studies, these empirical parameters are correlated to process performance and/or product quality. Unfortunately, due to the complex and non-ideal nature of these empirical tests, their results are difficult to interpret, and the parameters are not often physically meaningful.

The uniqueness of wheat is exemplified in breadmaking. During fermentation and baking, the ability of a dough to retain and stabilize the gas-filled bubbles and to prevent their premature rupture during processing and expansion is a property unique to wheat; the associated rheological and physical properties are important to the structure and eating quality of the bread loaf. As the bubbles grow, during proofing and baking, the cell walls stretch and become thinner. The molecular structures and network of the wheat flour polymers confer strength and stability to the thinning bubble walls, preventing premature rupture. This behavior is reflected in the unique properties of wheat dough rheology under these stretching conditions or during extension. Extensional rheological behavior is also important during dough sheeting and to gluten development during dough mixing. Thus, extensional test instruments are important in quantifying these unique properties of wheat dough.

Experimentally generating an extensional flow for rheological measurements is extremely difficult [4, 5]. Empirical extensional test instruments achieve rheologically non-ideal extensional flows. Amongst the types of extensional flows recognized in rheology [4, 5], the Extensograph and the Kiefer Extensibility rig generate uniaxial extension, while the Alveograph generates biaxial extension. The uniaxial extension in the Extensograph and the Kiefer Extensibility rig is achieved by stretching a nearly cylindrical sample of dough; in principle, this is similar to stretching a cylindrical sample by pulling it by its ends. The biaxial extension in the Alveograph is achieved by inflating a dough bubble, which is similar in principle to the mechanics of the expansion of gas bubbles during fermentation and baking. Biaxial extension may also be achieved using the lubricated squeeze flow technique [6–8]. The Alveograph is particularly important, since it is commonly used in the flour industry to characterize the biaxial extensional rheological behavior of wheat doughs, thereby making it technologically important.

The bubble inflation test, as with the Alveograph, has also been popular for rubbers [9], elastomers [10, 11], and polymers [12], in recognition of its many benefits as a rheological measurement of biaxial extensional behavior. All these studies and analyses were motivated toward fundamental rheological characterization and to developing the technique into a rheometer. Rheometrical tests tend to be near-ideal by design, and thus the measurements lend themselves to scientifically rigorous analysis, in order to obtain fundamental rheological material properties. Fundamental rheological properties allow strong scientific interpretation, can be related to material structure and molecular-level interactions between chemical constituents, and are directly applicable to processing, and to building rigorous mathematical models.

Despite the difficulty in analyzing the complex and non-ideal kinematics of the empirical Alveograph test, many attempts have been made to do so [13–18].

In this chapter, we build on that past research, in order to extract rheologically and physically meaningful parameters from the Alveograph test, thereby enhancing the power and value of the results for industrial practice. The objectives, therefore, are to

- Extend the analysis of the Alveograph test, in order to retrieve a simple and strong model for biaxial extensional viscosity,
- Illustrate the rheological and physical meaning of the parameters from the model, and
- Demonstrate the use of an Alveograph property map for the practically useful physical interpretation of the results, in order to benefit industrial practitioners.

II. ALVEOGRAPH TECHNIQUE

In 1905, J. von Hankoczy developed a bubble inflation test for gluten, by inflating a 2 cm diameter gluten sheet (Chapter 1 in Ref. [18]). Marcel Chopin, in the 1920s, independently designed and developed an "extensimeter", which later evolved to be commercialized as the Chopin Alveograph (Chapter 1 in Ref. [18]). The Chopin Alveograph has been a popular measurement technique used by the wheat flour industry globally for over a century, with standard methods (e.g., AACC 54-30.02 or 54-30A [19]) well established.

The method starts with the preparation of a disk of dough, using a standard protocol consisting of a standard formula, mixing, and handling protocol. The disk of dough, clamped around its perimeter, is inflated by pumping air at a constant flow rate, until the dough bubble bursts. The pressure, *P*, is recorded as a function of time (Figure A.1). All the results reported in this chapter were obtained by following the experimental protocol described in the AACC 54-30A standard method.

Several parameters may be extracted from the results ([1], Chapter 3 in Ref. [18]). In its simplest version, the peak pressure is given by *P*, representing "resistance" to deformation, the point of bubble rupture is given by the distance *L*, representing "extensibility", and the area under the measured curve is given by *W*, representing "strength".

There is no common agreement to date on the physical interpretation of the myriad parameters from the alveogram (Chapter 5 in Ref. [18]); nor is there any consensus on the relationship of the empirical parameters from the alveogram to critical functional flour properties (such as those due to the key wheat flour polymers, viz., gluten protein, pentosans, and damaged starch). Interpretations of the alveogram parameters and their correlations to flour quality and baked product properties are

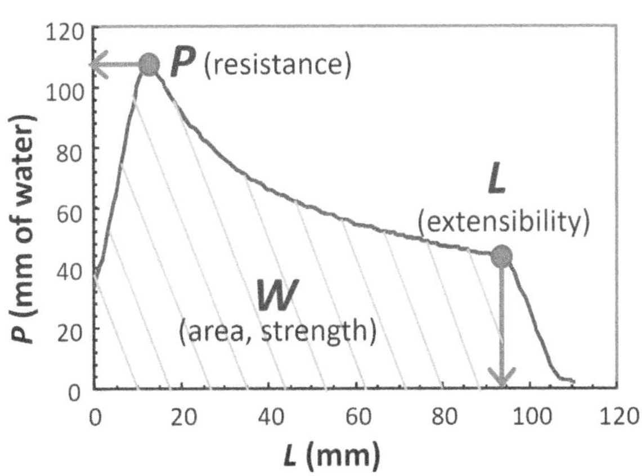

FIGURE A.1 Illustration of a typical alveogram. The measured pressure, *P*, is given on the *y*-axis. Since the data historically were recorded on a chart recorder, the *x*-axis is given in terms of a distance, *L*, and relates to time.

limited by the scope and results of each study (Chapter 5 in Ref. [18]). This has limited the value that the cereal scientist can derive from this unique and important rheological test for flour quality quantification. The analysis, parameters, and interpretation developed in this chapter should help to reveal the hidden rheological and practical value of the Alveograph test.

III. THEORY

A. Geometrical Analysis

A circular disc of dough, with an initial thickness δ_0 and radius a, is inflated under a constant volumetric flow rate, as illustrated in Figure A.2. Under ideal conditions, the dough wall thickness will decrease uniformly during inflation. In practice, the bubble will inflate nonuniformly.

Following Hlynka and Barth [13], the case of the bubble with its wall thickness thinning uniformly during inflation is considered. At any instant in time during

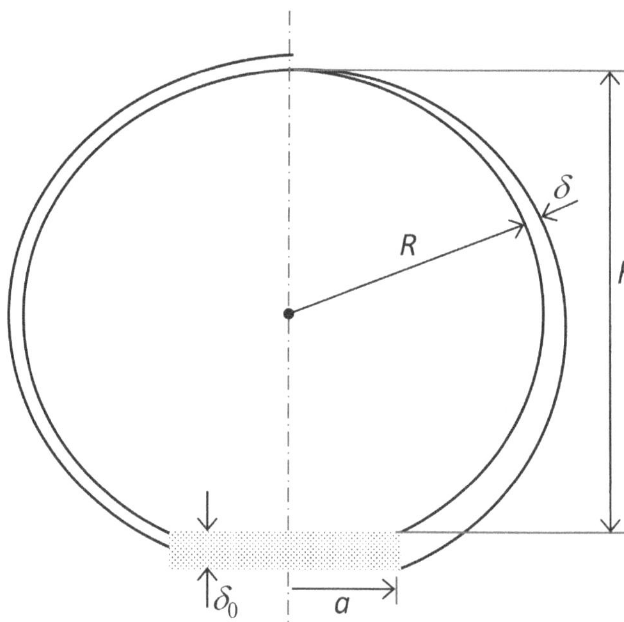

FIGURE A.2 Illustration of a circular disc of dough, of initial radius a (= 2.75 cm) and initial thickness δ_0 (= 0.25 cm), being inflated at a constant volumetric flow rate Q (= 27 cm^3/s), into a spherical bubble of radius $R(t)$, height $h(t)$, and a bubble wall thickness of $\delta(t)$, at an arbitrary time t during inflation. The left half of the diagram shows a uniform bubble wall thickness, while the right half shows the more realistic nonuniform bubble wall thickness during inflation. Bloksma [14] assumed the dough particles to move perpendicular to the surface of the bubble during inflation, in order to allow the calculation of the nonuniform dough wall thickness at any time and position; for example, an element of dough located at r would trace the path shown by the dashed line during inflation.

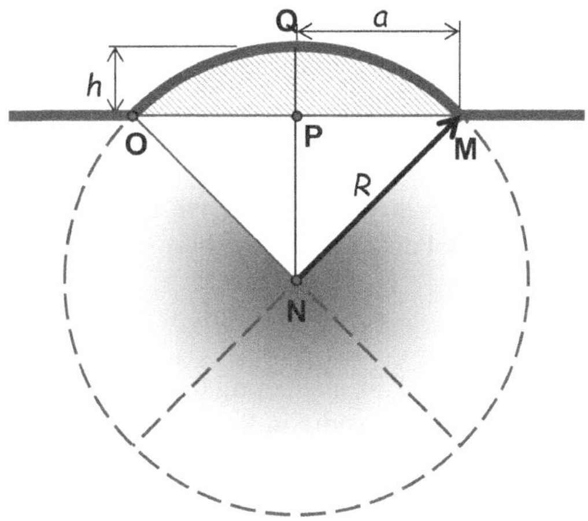

FIGURE A.3 Geometrical considerations for a circular disc of dough, lying along the line OPM and clamped around its perimeter at O and M, and inflated from the bottom, forming a bubble shaped as the spherical segment POQM at time *t*.

inflation, the dough bubble may be assumed to be shaped as a spherical segment (Figure A.3; [9]). The volume $V(t)$ of this spherical segment POQM is given by

$$V(t) = \frac{1}{6}\pi h(t)\left(3a^2 + h(t)^2\right)$$

Since the bubble is inflated at a constant volumetric flow rate Q, $V(t) = Qt$, resulting in a cubic equation for $h(t)$

$$h(t)^3 + 3a^2 h(t) - \frac{6}{\pi}Qt = 0 \quad \text{(A.1)}$$

This can be solved for the bubble height to give

$$\boxed{h(t) = \left[\frac{3Qt}{\pi} + \sqrt{a^6 + \left(\frac{3Qt}{\pi}\right)^2}\right]^{\frac{1}{3}} + \left[\frac{3Qt}{\pi} - \sqrt{a^6 + \left(\frac{3Qt}{\pi}\right)^2}\right]^{\frac{1}{3}}} \quad \text{(A.2)}$$

The surface area $A(t)$ is given by

$$A(t) = 2\pi R(t)h(t) \quad \text{(A.3)}$$

For the right-angled triangle NPM

$$R(t)^2 = a^2 + \left[R(t) - h(t)\right]^2$$

This gives

$$\boxed{R(t)=\frac{a^2+h(t)^2}{2h(t)}} \tag{A.4}$$

Assuming that the dough is incompressible, and the volume of the dough in the bubble is the same as its initial disc volume, we have

$$A(t)\delta(t)=A_0\delta_0$$

Inserting Eq. (A.4) into Eq. (A.3), and noting that the dough is initially a circular disc ($A_0=\pi a^2$), yields (see the Appendix in Ref. [14] for derivation of this equation from first principles)

$$\frac{\delta(t)}{\delta_0}=\left(\frac{a^2}{a^2+h(t)^2}\right) \tag{A.5}$$

Bloksma [14] observed the nonuniformity in the dough wall thickness during inflation and assumed the dough particles to move perpendicular to the bubble surface, in order to arrive at the following equation for the dough wall thickness at any position *r*:

$$\frac{\delta(t)}{\delta_0}=\left[\frac{a^4+r^2h(t)^2}{a^2\left(a^2+h(t)^2\right)}\right]^2$$

At the pole of the bubble (where $r=0$), it is generally recognized that near-ideal or equi-biaxial extension is achieved, the wall thickness changes uniformly, and Bloksma's general equation given above reduces to

$$\frac{\delta(t)}{\delta_0}=\left(\frac{a^2}{a^2+h(t)^2}\right)^2 \tag{A.6}$$

This equation can be generalized by introducing the parameter *c*, giving

$$\boxed{\frac{\delta(t)}{\delta_0}=\left(\frac{a^2}{a^2+h(t)^2}\right)^{2c}} \tag{A.7}$$

For $c=0.5$, the case of uniform wall thickness Eq. (A.5) is obtained, while $c=1$ gives Bloksma's nonuniform wall thickness case Eq. (A.6). With $h(t)$ given by Eq. (A.2), the dough bubble wall thickness can be predicted at any time of inflation.

Charalambides et al. [17, 20] demonstrated, through careful experimental measurements, that the dough wall thickness is indeed nonuniform during inflation, but it does not follow Bloksma's prediction Eq. (A.6). Analyzing their results yields a value of 0.85 for *c*. Remarkably, $c=0.85$ was obtained also by analyzing the results for elastomers [11].

It was therefore taken that the nonuniform change in bubble wall thickness during inflation is similar in most cases and can be expressed by Eq. (A.7), with $c = 0.85$.

B. Rheological Analysis

For an infinitesimally small time interval, the bubble wall thickness $\delta(t)$ will decrease by $d\delta$. The corresponding infinitesimal strain in the radial direction is

$$d\varepsilon_R = -\frac{d\delta}{\delta(t)} \tag{A.8}$$

The negative sign signifies that the thickness is decreasing. By assuming the dough to be incompressible, and using spherical coordinates, the strain in the direction of extension is

$$\varepsilon = \frac{1}{2}\varepsilon_R \tag{A.9}$$

Since the deformation is large in the Alveograph test, the Hencky measure of strain needs to be determined. By combining Eqs. (A.8) and (A.9), the Hencky strain ε_b is given by

$$\varepsilon_b = \int_{\delta_0}^{\delta(t)} -\frac{1}{2}\frac{d\delta}{\delta(t)}$$

Integrating this equation results in

$$\varepsilon_b = -\frac{1}{2}\ln\frac{\delta(t)}{\delta_0} = \frac{1}{2}\ln\frac{\delta_0}{\delta(t)}$$

Inserting Eq. (A.7) yields the following for the Hencky strain

$$\boxed{\varepsilon_b = c\ln\left(\frac{a^2 + h(t)^2}{a^2}\right)} \tag{A.10}$$

Differentiating Eq. (A.10) with respect to time gives the strain rate $\dot{\varepsilon}_b$ as

$$\dot{\varepsilon}_b = 2c\left(\frac{h(t)}{a^2 + h(t)^2}\right)\frac{dh}{dt} \tag{A.11}$$

From Eq. (A.1),

$$\frac{dh}{dt} = \frac{2}{\pi}\frac{Q}{\left[a^2 + h(t)^2\right]}$$

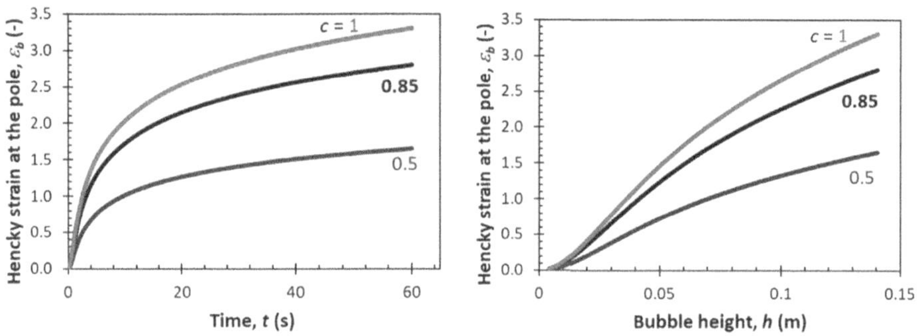

FIGURE A.4 Hencky strain from Eq. (A.10), as a function of (a) time and (b) bubble height, for different values of the empirical parameter c.

Therefore, the strain rate can be expressed by

$$\boxed{\dot{\varepsilon}_b = \frac{4c}{\pi}\frac{Qh(t)}{\left[a^2 + h(t)^2\right]^2}} \tag{A.12}$$

The strain and strain rate are plotted in Figures A.4 and A.5, respectively, as a function of inflation time and bubble height. When the bubble expands to achieve ε_b of 2, the disc of dough, with an initial thickness of 0.25 cm, thins down to a thickness of approximately 4.6×10^{-3} cm. Note that the strain rate is small throughout the test, reaching a maximum value (this can be calculated by differentiating Eq. (A.12) with respect to h, equating to 0, and solving) of about 0.5 s^{-1} at 0.78 s and dropping rapidly after that, for $c = 0.85$.

Next, the biaxial stress σ_b, experienced by the bubble at any time during inflation, is derived by applying a force balance in the y-direction and balancing against the measured pressure P, which acts uniformly inside the bubble and varies with time of inflation, as follows (Figure A.6):

$$\sigma_b A_{cs} \sin\theta = \pi a^2 P \tag{A.13}$$

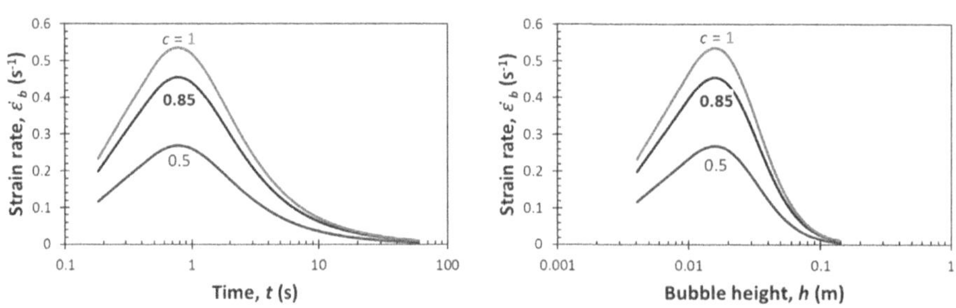

FIGURE A.5 Strain rate from Eq. (A.12), as a function of (a) time and (b) bubble height, for different values of the empirical parameter c. Note the logarithmic scale used for the x-axis.

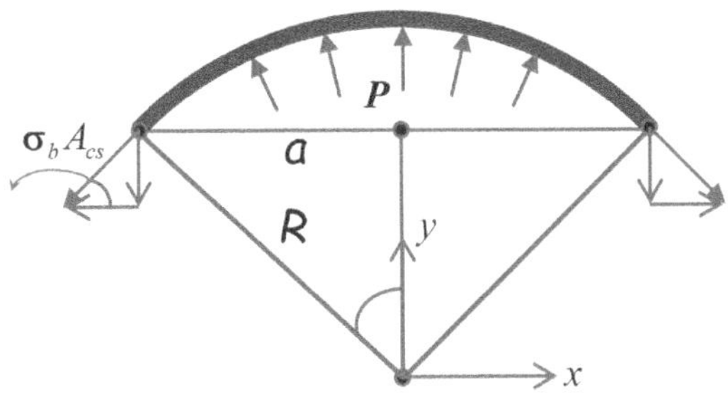

FIGURE A.6 Illustration of the tension on the dough film and the corresponding measured pressure P during inflation, in order to calculate the stress using force balance along the y-axis.

The cross-sectional area A_{cs}, over which the stress is acting, is given by

$$A_{cs} = 2\pi a\delta(t), \tag{A.14}$$

and

$$\sin\theta = \frac{a}{R(t)} \tag{A.15}$$

Note that both A_{cs} and $\sin\theta$ are changing during inflation, and hence are functions of time also.

Inserting Eqs. (A.14) and (A.15) in Eq. (A.13) and rearranging yields the following expression for the biaxial extensional stress:

$$\sigma_b(t) = \frac{PR(t)}{2\delta(t)} \tag{A.16}$$

Substituting Eqs. (A.4) and (A.7) into Eq. (A.16), the biaxial extensional stress can be given in terms of $h(t)$ as

$$\boxed{\sigma_b(t) = \frac{P}{4\delta_0}\frac{\left(a^2 + h(t)^2\right)^{2c+1}}{a^{4c}h(t)}} \tag{A.17}$$

Dividing the biaxial extensional stress in Eq. (A.17) by the strain rate in Eq. (A.12) gives the biaxial extensional viscosity η_b

$$\boxed{\eta_b = \frac{\pi}{16c}\frac{P}{Q\delta_0}\frac{\left(a^2 + h(t)^2\right)^{2c+3}}{a^{4c}h(t)^2}} \tag{A.18}$$

During fundamental rheological measurements, a constant strain rate, for example, is applied, until the measured stress reaches steady state; Yang and Dealy [12] designed a bubble inflation device with a control system, to perform as such a rheometer. Alternatively, a constant stress may be applied, until the strain rate reaches steady state [4, 5]. However, in the Alveograph, the bubble is inflated at a constant volumetric flow rate, and as already demonstrated, both the strain rate and the stress vary with inflation time Eqs. (A.12) and (A.17); consequently, the biaxial extensional viscosity is also a function of time.

Several studies ([15], Chapter 2 in [18, 21]) found that the dependence of the transient behavior of the biaxial extensional viscosity on strain and strain rate is separable and can be expressed as follows:

$$\eta_b = k e^{\alpha \varepsilon_b} \dot{\varepsilon}_b^{\,n-1} \tag{A.19}$$

where k is a fit coefficient (equivalent to η_b at $\varepsilon_b = 0$ and $\dot{\varepsilon}_b = 1$), α is a strain hardening index, and n is a power-law index. Studying a range of flours and test conditions, they found n to be in the range of 0.2–0.35 for $Q = 2.8$–$33\ \text{cm}^3/\text{s}$ and $V = 25$–$500\ \text{cm}^3$, indicating an extensional strain rate thinning behavior, as expected. For the standard Alveograph operating condition of $Q = 27\ \text{cm}^3/\text{s}$, the reported range of V [14] translates to $t = 0.95$–20 s. In Figure A5a $t = 0.95$ s is past the peak strain rate; the strain rate peaks at $t = 0.78$ s.

In practice, is η_b from the measurement a function of both ε_b and $\dot{\varepsilon}_b$?

Using the data at $t > 0.78$ s for $c = 0.85$, and analyzing η_b versus ε_b data for a wide range of flours, allowed simplification of Eq. (A.19), by removing the strain rate dependence, to obtain the following equation (Figure A.7):

$$\boxed{\eta_b = K e^{\alpha \varepsilon_b}} \tag{A.20}$$

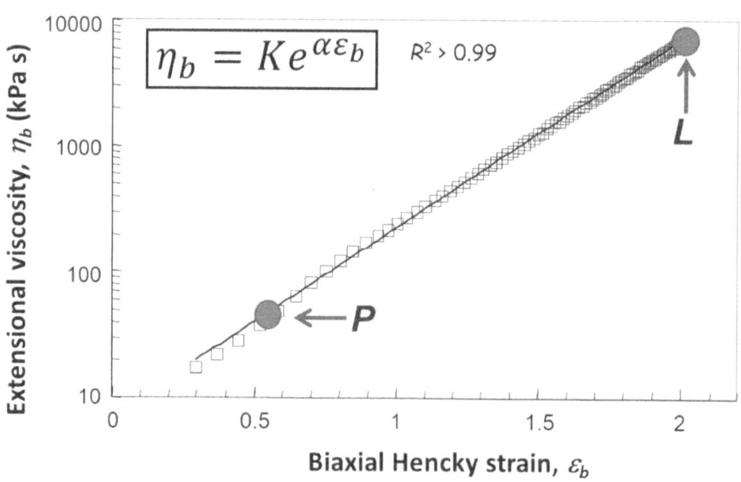

FIGURE A.7 Illustration of the strong fit to the experimental data from Alveography obtained using Eq. (A.20); in general, Eq. (A.20) fits Alveograph data, with an $R^2 \geq 0.95$, in a vast majority of cases for a wide range of flours. This fit allows the quantification of the entire measured P vs t curve, using just two unambiguous and physically meaningful fit parameters, K and α. K is the zero-strain extensional viscosity, with the units of Pa s, and α is the strain hardening index, with no units. Note the location of P and L on this curve as singular points.

where K is the zero-strain extensional viscosity (with the units of Pa s), and α is the strain hardening index (dimensionless). Over 100 flour samples were analyzed, resulting in an excellent fit of Eq. (A.20) ($R^2 > 0.95$, with the average error in the magnitudes of the parameters of 0.3 (i.e., ±0.15) kPa s for K and 0.1 (i.e., ±0.05) for α).

With experimental Alveograph data obeying Eq. (A.20) for such a large number and range of flours, it follows that the dependence of η_b on $\dot{\varepsilon}_b$ is negligible for the measurement! Two factors may explain this. First, dough may behave like polymers, with the viscosity being independent of the strain rate at low strain rates [22]; for the Alveograph test, as illustrated in Figure A.5, the strain rates do not exceed 0.5 s^{-1} and rapidly drop in magnitude, as the bubble is inflated. Second, the viscosity contribution, due to dough's strong strain hardening behavior ($\alpha > 2$ for a wide range of wheat doughs, consistently, as shown later in this chapter), may be dominating, thereby making the strain rate dependence negligible. The consistent validity of the model in Eq. (A.20), to a wide range of flours and under various conditions studied (see below), suggests that, in practice, and under the Alveograph test conditions, assuming the biaxial extensional viscosity to be a function of only the biaxial strain is reasonable.

Nevertheless, it is important to emphasize that the model represented by Eq. (A.20) captures the rheological behavior of the entire measured alveogram, with just two physically meaningful parameters. In principle and in practice, this is considerably superior to the traditional empirical parameters of P, L, W, etc. Furthermore, K and α capture two physically important and independent rheological properties of the wheat dough. The contributions of flour components, water content, and other ingredients to the viscosity are captured by K, while the contributions to bubble formation, growth, and stability due to strain hardening are captured by α. Functional and rheological significance and interpretation of K and α are discussed in detail in the following sections.

IV. LINKING *K* AND *α* TO WHEAT FLOUR POLYMERS

In the context of biscuit-making, the much greater importance of functional specifications over the more common compositional specifications for wheat flour has been emphasized [23]; the same is true for breadmaking and for most wheat-based products, in general. Flour polymers—gluten protein, pentosans, and damaged starch—are the three major functional components of wheat flour. These have been discussed in detail [23] and related to the traditional Alveograph empirical parameters of P, L, and W, through the SRC test [24].

The SRC test quantitatively assesses the individual functional contribution of each of the functional flour polymers [24]. This is done by choosing solvents that selectively solvate maximally each of the flour polymers. Dilute aqueous lactic acid solution is used for glutenins, dilute aqueous sodium carbonate (Na_2CO_3) solution for damaged starch, and concentrated aqueous sucrose solution for pentosans. In addition, water is used as the reference solvent; note that each of the other solvents is also aqueous. SRC is an AACC-approved standard method (AACC 56-11.02 [19]), which provides details of the test protocol. An excellent description of the history,

TABLE A.1
Flour Blends and Their Properties, Used to Determine the Relationship Between the Alveograph Rheological Parameters and the Functional Flour Components. All Flours Used Were of the Hard Red Variety

Flour ID	Wheat Blend	Protein (%)	Ash (%)	K (kPa s)	α
F1	60% spring, 40% winter	11	0.51	4.89	3.27
F2	100% spring	13	0.55	7.47	3.32
F3	100% spring	13.6	0.56	7.32	3.38
F4	100% spring	14.2	0.55	7.06	3.41
F5	100% winter	11.3	0.54	5.40	3.28
F6	80% spring, 20% winter	12.2	0.55	7.20	3.26
F7	40% spring, 60% winter	10.6	0.5	6.24	3.24
F8	100% spring	12.7	0.55	5.95	3.36
F9	80% spring, 20% winter	11.6	0.5	5.92	3.30
F10	100% spring	12	0.39	6.41	3.55
F11	50% spring, 50% winter	13.2	1.1	4.92	2.82
F12	50% spring, 50% winter	12.4	0.51	5.56	3.50

principles, applications, interpretation, and practical uses of the SRC method can be found in a recent review [24].

To determine the relationship between the new Alveograph rheological parameters, K and α, and the functional components of wheat flour, a study of 12 different flours was conducted (Table A.1). SRC measurements were also conducted on all these flours.

The results are summarized in Figure A.8. The zero-strain extensional viscosity K is strongly related to the SRC-Na_2CO_3 value for the damaged starch, and to the SRC-sucrose value for the pentosans. The strain hardening index α is strongly related to flour protein content and to the SRC-lactic value for the glutenins.

These results are in agreement with polymer rheology fundamentals [4]. Polymer viscosity is proportional to the cube of polymer molecular weight, approximately. Thus, a higher molecular weight and a higher polymer concentration will increase the viscosity. Molecular weights for wheat starch and pentosans are on the order of 10^5–10^9 [25] and 10^5 [26], respectively, and 10^5–10^7 or less for native glutenins [27]. Thus, increasing amounts of the flour polymers with much higher molecular weights (viz., damaged starch and pentosans) will more dominantly increase K, the zero-strain extensional viscosity. Long-chain branches lead to extension thickening and strain hardening in synthetic polymers [4]; a similar effect can be expected for flour polymers. During dough mixing, glutenins can form disulfide bonds, resulting in a network structure (and much higher molecular weights, 10^6–10^7). Higher amounts of glutenins, as detected by increased SRC-lactic values, will increase its network-forming ability in the dough. Thus, the strain hardening index α detects the extent of network formation in the dough; a stronger network leads to higher α, and a greater effect on baking expansion. Note that a small change in α has a large strain hardening effect, since η_b is an exponential function of α.

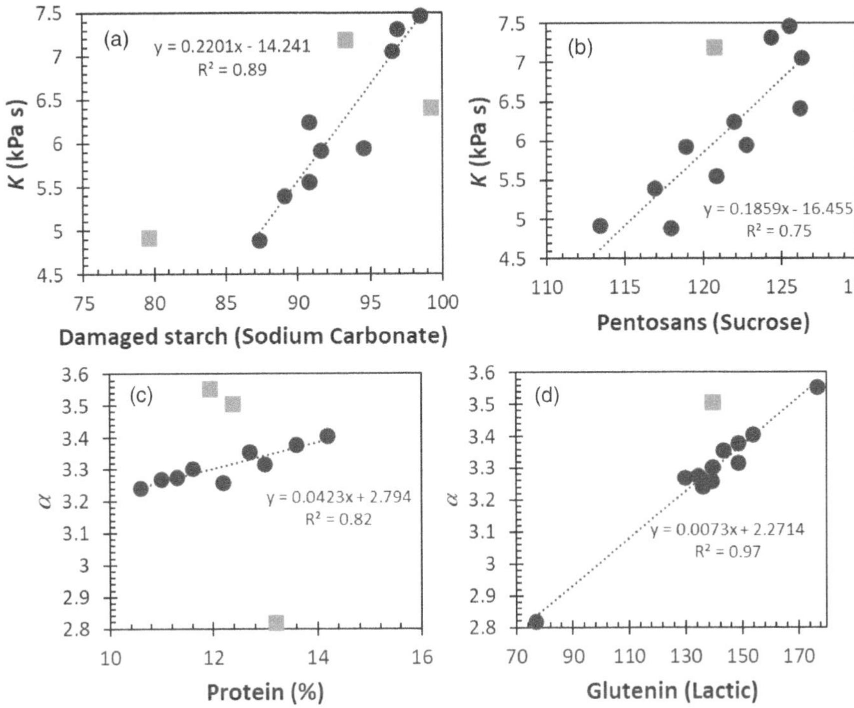

FIGURE A.8 (a–d) Relationship between the Alveograph K and α parameters and the functional flour polymers, through the SRC results and flour protein content. The linear fits are for the data (filled circles) excluding the outliers (filled squares).

V. THE ALVEOGRAPH PROPERTY MAP

To the industrial practitioner, the Alveograph property map (shown in Figure A.9) of K and α represents simple and practically useful format for rapidly deriving insights and making decisions.

Higher magnitudes of K indicate greater amounts of damaged starch and pentosans in the flour, loosely relating to the common wheat technology terminology of "soft" and "hard", as shown in Figure A.9. Higher magnitudes of α indicate more functional *glutenins* in the flour, loosely relating to the common wheat technology terminology of "weak" and "strong" (Figure A.9).

VI. INTERPRETATION OF K AND α

Technologically, dough viscosity is important to processing (mixing, proofing, forming, and baking) and final product quality (e.g., bread loaf volume and texture). By quantifying this property with a singular number, K, its use to industrial practitioners is greatly simplified. We have already seen that extensional dough rheology is important to baked goods (as in proofing, sheeting, and baking); hence, K being an extensional viscosity parameter is physically meaningful and practically important.

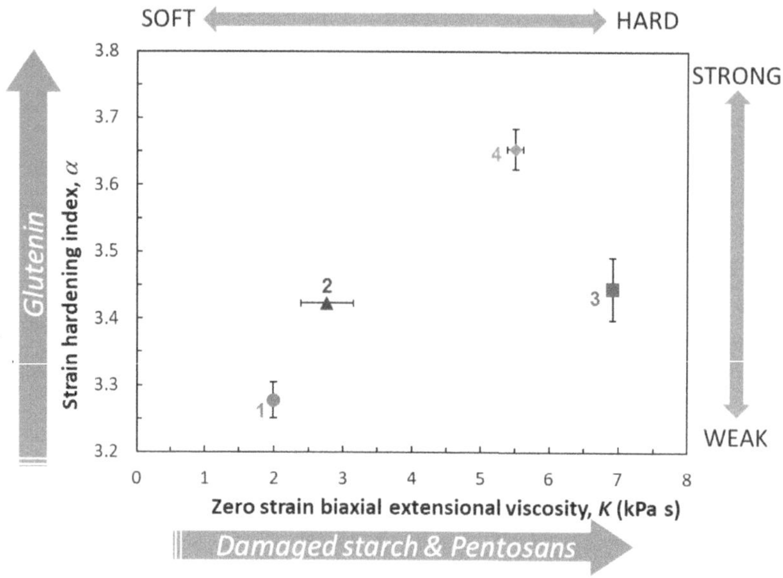

FIGURE A.9 Alveograph property map, with K on the abscissa and α on the ordinate. The numbers next to the data points indicate different flours/flour blends: 1—100% soft red winter, 2—90:10 soft:hard red winter blend, 3—50:50 soft:hard red winter blend, and 4—100% hard red spring.

Additionally, shear viscosity is also practically important. In rheological literature, it is well-known that the extensional viscosity is typically many times the magnitude of shear viscosity, approaching a constant integer multiple in the limit of low strains; for the biaxial extensional viscosity, this is six times the shear viscosity [4]. Thus, K quantifies both shear and biaxial extensional viscosity in practice.

The rheological phenomenon of strain hardening confers stability to the growing bubble, by strengthening the thinning bubble wall as the bubble expands. This delays the bubble rupture during proofing and baking, as well as during sheeting; it also manifests as elasticity of the dough and as the commonly observed "snap-back" behavior. The technological importance of strain hardening is well recognized in polymer applications and is linked to molecular structure [4, 28, 29]; it is noteworthy that molecular structures dominating the extensional rheology, especially strain hardening, are not all the same as those dominating the shear rheology [28]. The property of wheat dough which permits the retention of gas during baking has been variously known as "extensibility" and "elasticity". In a seminal paper, van Vliet et al. [30] demonstrated that the rheological strain hardening is the quantitative measure that confers stability to growing bubbles in dough, thus allowing gas retention; they also derived stability criteria for growing bubbles, based on rheology, using previously published lubricated squeeze flow test results. In a subsequent review, van Vliet [31] demonstrated that those stability criteria remain valid for bubble inflation tests such as the Chopin Alveograph, provided that the non-constant strain rate, prevalent during the test, is taken into consideration. With $\alpha > 2$ for all dough samples studied to date, it can be concluded that wheat doughs, in general, are inherently stable to bubble inflation. It is, therefore, the degree of stability that differentiates among flours, thus making certain flours more suitable for certain products (such as breads). Eventually, any bubble will rupture, when the thinning bubble wall can no longer resist the bubble pressure.

A third measure, which is considered important in several published studies on dough, is the point of bubble rupture—given by the L value of the alveogram (Figure A.1), or alternatively as elongation or strain at rupture. In the case of bread, for example, rupture of bubbles in the dough (during proofing) clearly will be detrimental to expansion of dough during baking. However, rupture and coalescence of bubbles occur during baking, with the final bread loaf consisting of both closed and open cells; therefore, relating the dynamics of cell rupture during baking to the strain at rupture, measured by Alveography on the dough, is mechanistically complex and difficult to analyze and predict. Additionally, L is known to be highly variable (see Chapter 5 in Ref. [18]), due to heterogeneity of the dough and related defects, among other factors; during baking, there are additional factors (such as temperature-induced physical and chemical changes, the population of bubbles, and the interactions among them) that introduce randomness. Thus, strain at rupture, measured using the Alveograph (or L), is dominated by statistics, with questionable correlation to final cell structure of the baked product, such as bread, and is therefore not a physically useful quantity.

Since the Alveograph test is ordinarily conducted on a dough made of wheat flour and salt water, at room temperature, the results (K and α) will have the strongest quantitative relationships to flour quality and processability of the dough (such as mixing, proofing, and forming operations); those relationships are maintained to the extent that the effects of mixing conditions and formulations (for ingredients that modify dough rheology) can be measured and quantified using the Alveograph test. The physics-based relationships of these parameters to baking performance are likely to be weaker, and even more so to the final baked product properties. However, controlled empirical studies may help to establish practically useful correlations, which may be strengthened by using well-designed statistical and data science-based analysis and models.

VII. APPLICATIONS OF THE NEW ANALYSIS AND THE PROPERTY MAP

To deliver functional flours with consistent properties over multiple batches and crop years, it is common practice amongst flour millers to blend flours. In some cases, flour millers are known to use the Alveograph P-value (Figure A.1), aiming to deliver consistent flour blends. In practice, this has not always been reliable; the analysis presented in this chapter demonstrates the reason for the lack of reliability of P. As illustrated in Figure A.10, flour blends formulated to achieve a constant P-value can have wide ranges of K and α values and thus will exhibit widely different rheological behaviors. Consequently, a baker often needs to do additional time-consuming trial-and-error adjustments to the dough composition and mixing conditions, in order to deliver the desired dough rheology out of the mixer and thereby maintain consistent consumer-desired eating quality of products.

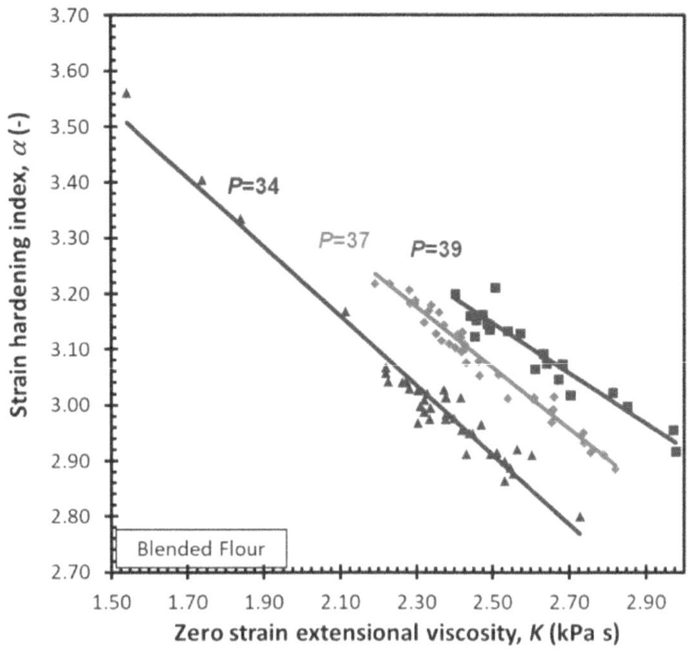

FIGURE A.10 Alveograph property map illustrates that several flour blends, formulated to specific target *P*-values, are rheologically not same.

Instead, by using the K and α Alveograph parameters, a flour miller can establish a two-dimensional target space on the Alveograph property map and blend flours to deliver consistent rheology to customers. Figure A.11 illustrates the results of a study done during seasonal crop change, when flours were blended using the *P*-value, but through a trial-and-error process over two months. The filled circle symbols represent the target specifications on the Alveograph property map, for flour blends from the previous crop, which produced dough batches that processed well. Each open symbol represents one blend ratio; thus, 40 blend ratios and batches of flour and dough were tested over two months, in order to arrive at an acceptable flour blend. Thus, the Alveograph property map can serve as a useful and convenient tool to the industrial practitioner, to define and arrive rapidly at the target dough rheology when blending flours, at a much-reduced cost, thereby ensuring consistency. Using an appropriate viscosity blending rule [32] would further enhance this exercise.

Commercial formulations of doughs used to manufacture baked goods typically consist of several ingredients, in addition to flour, water, and salt [33]. Many of these additional ingredients can alter the dough rheology, with some deliberately added to effect changes to the dough rheology for improved processability, while others are intended to help achieve a desired set of product properties, such as structure, texture, quality, and shelf-stability. In addition, the dough mixing time and mixing conditions can affect the dough rheology [34]. Thus, a quantitative understanding of

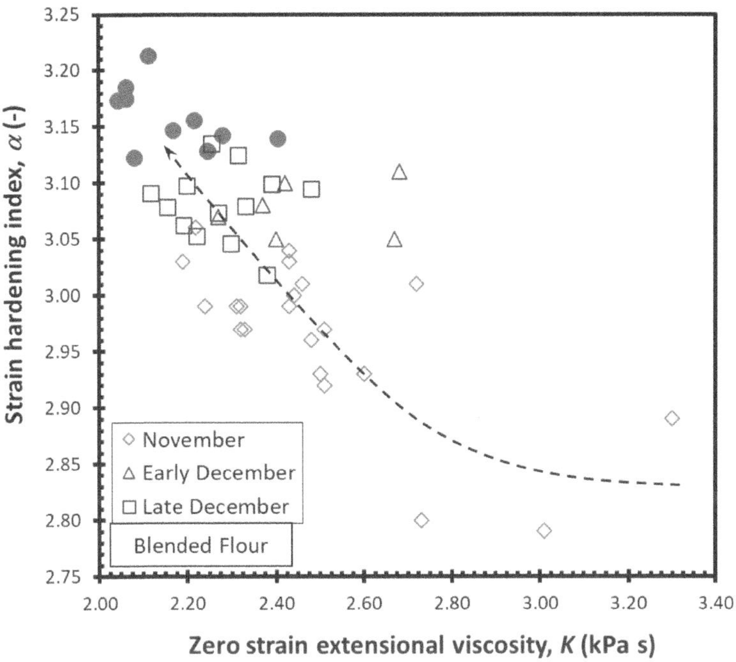

FIGURE A.11 Traditional approach to blending flours, in order to ensure consistency over time (such as crop change), is expensive and time-consuming. The filled circles represent the desired dough rheology of the flour blend from the prior year crop, while the open symbols represent the trial-and-error approach used to produce multiple flour blends for the new crop. The property map, using the new Alveograph analysis, offers a simple and more useful approach to rapidly arrive at the desired dough rheology.

the effect of ingredients and mixing is beneficial, in order to better control the post-mixing manufacturing process and thereby deliver consistent product quality. Akin to principles applied in the chemical industry, the added ingredients to the dough and the mixing conditions may be viewed as modifiers of the flour polymer properties. Thus, Alveograph measurements, using the new analysis and visualized through the property map, can provide a quantification of the effects of mixing and added ingredients on dough rheology.

Figures A.12 and A.13 illustrate the effects of dough mixing time and additives, respectively, on the Alveograph property map. In the mixing example (Figure A.12), mixing times are found to generally have a much larger effect on K (i.e., on polymer solvation), except for a narrow range of mixing times around 4 min, where α (i.e., glutenin network formation) may be maximized. The drop in α for longer mixing times may be interpreted either as “unmixing” [35], although in this case, the mixing speed has been constant for all mixing times, or as the breakdown of the glutenin network (conceptually, this may be interpreted as either network scission or disentanglement of the glutenin polymers), due to overmixing [36].

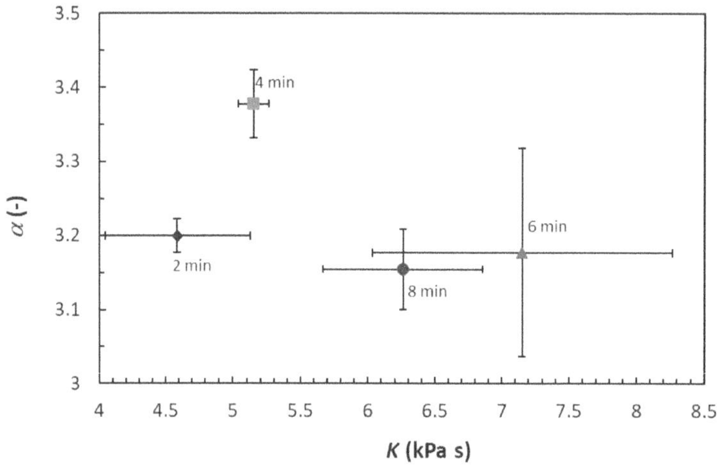

FIGURE A.12 Effect of dough mixing time on the Alveograph property map.

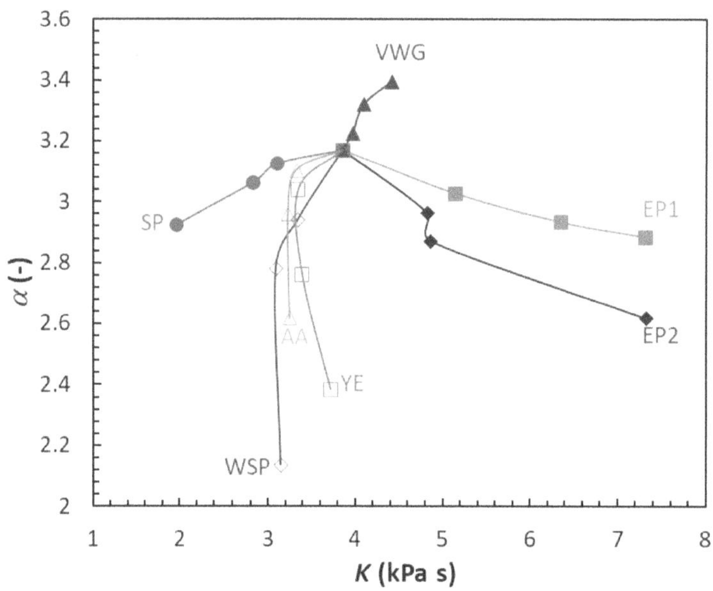

FIGURE A.13 Effects of additives on Alveograph property map.

To the product developer, a vast range of ingredients is offered by ingredient manufacturers, in order to modify dough rheology (especially by affecting the flour polymers) and thereby deliver various benefits. Often, due to the lack of effective measurement techniques, for use by ingredient suppliers and product developers, a clear quantitative understanding of the modifying behavior remains elusive, leading

TABLE A.2
Additives Used to Modify Functional Dough Rheology: WSP—Water-Soluble Protein, VWG—Vital Wheat Gluten, AA—Ascorbic Acid, and YE—Yeast Extract

Label	Effect	Amount
EP1	Emulsifying protein	1, 2, 5% (w/w)
EP2	Emulsifying protein	
SP	Softening protein	
WSP	Reducing elasticity	
VWG	Extensibility	
AA	Softener	10, 20, 50 ppm
YE	Softener	0.025–0.1%

to multiple trial-and-error studies and subjective assessments. Carefully chosen measurements, using the Alveograph property map, can be useful. This is illustrated in Figure A.13 for a selected set of additives (Table A.2).

Each of the additives and their amounts used in the dough are listed in Table A.2. Column 2 in this table also notes the purported function of each additive in the dough. The emulsifying and softening proteins strongly affect K, by either increasing (EP's) or decreasing (SP) dough viscosity (Figure A.13). The rest of the listed additives affect α, by either decreasing (WSP, YE, and AA) or increasing (VWG) the strain hardening behavior. It is noteworthy that AA and YE weaken the glutenin network (i.e., decreased dough strength) but don't soften the dough, while VWG strengthens the glutenin network (which would make the dough less extensible). The effect of each additive can also be quantified, allowing the formulator to optimize function and cost.

VIII. DISCUSSION

Complexities of dough structure and rheology make dough a fascinating study material. The extent to which dough rheology is important for practical applications has been the subject of debate [35, 37]. There is broad agreement that biaxial extensional rheology of a dough, as exemplified by the magnitude of the extensional viscosity and especially by dough's strain hardening behavior, is important in breadmaking—during the proofing and baking stages—by conferring the unique and desirable properties of gas retention and expansion of the gas cells without rupture and collapse. These rheological properties are also important to the manufacture of other bread-like products such as pizza. With the application of the new analysis, the traditional Alveograph test can be converted into a practically useful rheological test (even if non-ideal).

Various stress–strain equations have been used in the literature. Some studies [15, 30, 31] used $\sigma = ke^{\alpha\varepsilon_b}\dot{\varepsilon}^n$, which indicates that dough is considered a *liquid*. Others have used equations such as $\sigma = K\varepsilon^n$ [16] and $\sigma = Ke^{\alpha\varepsilon_b}$ [38], indicating that dough is considered a *solid*. None of these studies reported the quality of the fit of the equations used. In the present study, a simpler equation for stress, $\sigma = ke^{\alpha\varepsilon_b}\dot{\varepsilon}$, which also indicates that dough is considered a *liquid*, fitted the Alveograph data with excellent statistical confidence ($R^2 > 0.95$ for a wide range of flour samples). The dough composition and its processing determine the nature of the dough response as a liquid or solid, and this has also been the subject of debate [39].

Much has been made of the rupture point of the bubble in the Alveograph test. As a theoretical exercise, it may be of interest. However, as already noted earlier, there is too much "noise" (i.e., variability) in the rupture point, whether quantified as a failure strain or through the L-value at failure in the conventional P versus L curve, for it to be of any use in offering fundamental insights or for practical applications. Some [16, 30, 31] have applied the Considère construction to predict the stability, necking, and rupture or failure of a dough in extension. For viscoelastic materials such as synthetic polymers (dough also being viscoelastic), the applicability of the Considère criterion (developed for solid, and hence elastic, materials) has been questioned [40], even for uniform extensional flows. The nonuniform nature of the Alveograph test, and of bubble inflation during fermentation and baking, raises further doubts on the applicability of the Considère criterion for bubble rupture. Grenier et al. [41] have described four possible situations, based on dough structure and rheological factors, for the rupture of gas cells during bread baking.

During fermentation and baking, both extensional rheology and interfacial properties are important to bubble expansion, stability, and coalescence [35, 37, 41]. Overall dough expansion is achieved by the growth of many bubbles, their interactions, and coalescence. The mechanics and conditions are very different from the near-ideal Alveograph test conditions for a single bubble. It would be useful to assess the importance and therefore the predictability of the Alveograph test, by rigorously determining statistical correlations of K and α to process performance and product properties for baked goods.

Most significantly, the new analysis and the property map of K and α provide a more robust and scientifically rigorous quantitative approach to applying the Alveograph test for wheat breeding, flour consistency, setting specifications for different applications, and for the effect of selected additives and mixing conditions, in order to guide product and process development in the baked goods industry.

IX. CONCLUSIONS

Alveography is a powerful and practically relevant technique used to measure the unique extensional rheological behavior of wheat dough in biaxial mode. It serves to quantify the flour functional rheological properties critical to dough processing and baking performance, and thus to final products. The new and original analysis presented in this chapter elicits two quantitative rheological measures, viz., the zero-strain extensional viscosity K and the strain hardening index α, from the standard Alveograph test. Using SRC results, K and α are shown to be sensitive to the major functional flour polymers, K to damaged starch and pentosans, and α to glutenin. The results are presented in the form of a property map that represents a simple and practically useful format to rapidly derive insights and make decisions by the industrial practitioner. Thus, the analysis provides a standardized rheologically rigorous protocol, with output parameters that have both rheological and physical meaning, for use by research scientists and industrial practitioners. To a flour miller, this approach provides a rapid and cost-effective means to deliver consistent flour quality tuned to specific applications. Together with SRC, the Alveograph property map of K and α can serve as part of a powerful set of tools for wheat breeding. In addition to using the standard AACC protocol, from the dough preparation through the inflation step, for the Alveograph test, one can also envision using different formulations (especially involving functional additives), different mixing conditions, different mixers [42], and different proofing conditions, while using the standard inflation conditions of the Alveograph, in order to derive greater quantitative understanding and insights to support industrial formulations and processing of dough-based products.

SRC and Alveography are two complementary and powerful techniques, based on scientifically rigorous principles, for the characterization of wheat flour functionality. The new analysis of Alveograph data, presented in the form of its property map, has revealed the underlying rheological measures, thereby allowing the extraction of theoretically and practically important quantitative properties from this test, in order to benefit cereal researchers and industrial practitioners. Together, these tools can now enable wheat breeders, growers, and millers to deliver functionally more consistent flour quality, and enable product and process developers and bakers to manufacture products with consistently enhanced quality.

ACKNOWLEDGMENTS

Permission by the management of Mondelēz International to publish this work is gratefully acknowledged. MP's many hours of discussion with Drs. Louise Slade and Harry Levine on Food Polymer Science, flour functionality, and the enigma of conventional Alveograph results served as motivation for this work. Diane Gannon's and Brent Gradishar's collaboration and tutorials on the intricacies of the Alveograph measurements were critical to ensuring that the approach developed in this appendix is tailored to industrial practice. MP is grateful to Dr. Harry Levine's personal attention to reviewing and editing this appendix. Professor Rakesh Gupta's (West Virginia University) advice and encouragement are also gratefully acknowledged.

NOTE

1. Coauthored with Yongwoo Inn, formerly with Kraft Foods (Mondelēz International).

REFERENCES

1. Bloksma, A.H., and Bushuk, W. 1988. Rheology and chemistry of dough. Chapter 4, pages 131–217 in: *Wheat: Chemistry and Technology*, 3rd edn., Y. Pomeranz, Ed., American Association of Cereal Chemists, St. Paul.

2. Weipert, D. 2006. Rheology. Chapter 13.1, pages 117–146 in: *Future of Flour: A Compendium of Flour Improvement*, L. Popper, W. Schäfer, and W. Freund, Eds., ISBN 978-3-86037-309-5, Erlingen Verlag (Agrimedia GmbH), Clenze.
3. Hadnađev, T.D., Pojić, M., Hadnađev, M., and Torbica, A. 2011. The role of empirical rheology in flour quality control. Chapter 18, pages 335–360 in: *Wide Spectra of Quality Control*, I. Akyar, Ed., ISBN 978-953-307-683-6, InTech, London.
4. Dealy, J.M., and Wissbrun, K.F. 1990. *Melt Rheology and its Role in Plastics Processing: Theory and Applications*, ISBN 0-442-22099-5, Van Nostrand Reinhold, New York.
5. Macosko, C.W. 1994. *Rheology: Principles, Measurements, and Applications*, ISBN 1-56081-579-5, VCH, New York.
6. Casiraghi, E.M., Bagley, E.B., and Christianson, D.D. 1985. Behavior of mozzarella, cheddar, and processed cheese spread in lubricated and bonded uniaxial compression. *J. Texture Stud.*, 16:281.
7. Chatraei, S.H., Macosko, C.W., and Winter, H.H. 1981. Lubricated squeezing flow: A new biaxial extensional rheometer. *J. Rheol.*, 25:433.
8. Janssen, A.M., van Vliet, T., and Vereijken, J.M. 1996. Fundamental and empirical rheological behaviour of wheat flour doughs and comparison with breadmaking performance. *J. Cereal Sci.*, 23:43.
9. Rivlin, R.S., and Saunders, D.W. 1951. Large elastic deformations of isotropic materials. VII. Experiments on the deformation of rubber. *Phil. Trans. Roy. Soc. London, Ser. A*, 243:251.
10. Joye, D.D., Poehlein, G.W., and Denson, C.D. 1972. A bubble inflation technique for the measurement of viscoelastic properties in equal biaxial extensional flow. *Trans. Soc. Rheol.*, 16:421.
11. Reuge, N., Schmidt, F.M., Le Maoult, Y., Rachik, M., and Abbe, F. 2001. Elastomer biaxial characterization using bubble inflation technique. I: Experimental investigations. *Polym. Eng. Sci.*, 41:522.
12. Yang, M.-C., and Dealy, J.M. 1987. Control of strain rate in a sheet-inflation rheometer. *J. Rheol.*, 31:113.
13. Hlynka, I., and Barth, F.W. 1955. Chopin Alveograph studies. 1. Dough resistance at constant sample deformation. *Cereal Chem.*, 32:463.
14. Bloksma, A.H. 1957. A calculation of the shape of the alveograms of some rheological model substances. *Cereal Chem.*, 34:126.
15. Launay, B. and Buré, J. 1977. Use of the Chopin Alveographe as a rheological tool. II. Dough properties in biaxial extension. *Cereal Chem.*, 54:1152.
16. Dobraszczyk, B. J., and Roberts, C.A. 1994. Strain hardening and dough gas cell-wall failure in biaxial extension. *J. Cereal Sci.*, 20:265.
17. Charalambides, M.N., Wanigasooriya, L., Williams, J.G., Goh, S.M., and Chakrabarti, S. 2006. Large deformation extensional rheology of bread dough. *Rheol. Acta*, 46:239.
18. Dubois, M., Dubat, A., and Launay, B. 2008. *The AlveoConsistograph Handbook*, 2nd edn., ISBN 978-1-891127-56-4, AACC International, St. Paul.
19. AACC International. 2010. Methods 54-30.02, 54-30A, and %6-11.02 in: *AACC Approved Methods of Analysis*, 11th edn., Available online only, AACC International, St. Paul.
20. Charalambides, M.N., Wanigasooriya, L., Williams, J.G., and Chakrabarti, S. 2002. Biaxial deformation of dough using the bubble inflation technique. I. Experimental. *Rheol. Acta*, 41:532.
21. McCann, T.H., Le Gall, M., and Day, L. 2016. Extensional dough rheology—Impact of flour composition and extension speed. *J. Cereal Sci.*, 69:228.
22. Bird, R.B., Armstrong, R.C., and Hassager, O. 1987. *Dynamics of Polymeric Liquids, Volume 1: Fluid Mechanics*, 2nd edn., ISBN 0-471-80245-X (v. 1), Wiley, New York.

23. Kweon, M., Slade, L., Levine, H., and Gannon, D. 2014. Cookie vs cracker-baking—What's the difference? Flour functionality requirements explored by SRC and alveography. *Crit. Rev. Food Sci. Nutr.*, 54:115.
24. Kweon, M., Slade, L., and Levine, H. 2011. Solvent retention capacity (SRC) testing of wheat flour: Principles and value in predicting flour functionality in different wheat-based food processes and in wheat breeding—A review. *Cereal Chem.*, 88:537.
25. Tester, R.F., Karkalas, J., and Qi, X. 2004. Starch—Composition, fine structure and architecture. *J. Cereal Sci.*, 39:151.
26. Courtin, C.M., and Delcour, J.A. 2002. Arabinoxylans and endoxylanases in wheat flour bread-making. *J. Cereal Sci.*, 35:225.
27. Wieser, H. 2007. Chemistry of gluten proteins. *Food Microbiol.*, 24:115.
28. Graessley, W.W. 2008. *Polymeric Liquids and Networks: Dynamics and Rheology*, ISBN 0-8153-4171-7, Garland Science, New York.
29. Dealy, J.M., Read, D.J., and Larson, R.G. 2018. *Structure and Rheology of Polymers: From Structure to Flow Behavior and Back Again*, 2nd edn., ISBN: 978-1-56990-611-8, Hanser, Cincinnati.
30. van Vliet, T., Janssen, A.M., Bloksma, A.H., and Walstra, P. 1992. Strain hardening of dough as a requirement for gas retention. *J. Texture Stud.*, 23:439.
31. van Vliet, T. 2008. Strain hardening as an indicator of bread-making performance: A review with discussion. *J. Cereal Sci.*, 48:1.
32. Nielsen, L.E. 1978. *Predicting the Properties of Mixtures: Mixture Rules in Science and Engineering*, ISBN 0-8247-6690-3, Marcel Dekker, New York.
33. Faridi, H., and Faubion, J.M. 1990. *Dough Rheology and Baked Product Texture*, ISBN 0-442-31796-4, Van Nostrand Reinhold, New York.
34. Wrigley, C.W., Bekes, F., and Bushuk, W. 2006. Gluten: A balance of gliadin and glutenin. Chapter 1, pages 3–32 in: *Gliadin and Glutenin: The Unique Balance of Wheat Quality*, C. Wrigley, F. Bekes, and W. Bushuk, Eds., ISBN: 978-1-891127-51-9, AACC International, St. Paul.
35. MacRitchie, F.R. 2014. Requirements for a test to evaluate bread-making performance. *J. Cereal Sci.*, 59:1.
36. Meerts, M. 2018. *Assessing the role of the main flour components and fermentation in the rheology of wheat flour dough with a new rheological toolbox*, Doctoral dissertation, Faculty of Engineering Science, KU Leuven, Belgium.
37. Bloksma, A.H. 1990. Rheology of the breadmaking process. *Cereal Foods World*, 35:228.
38. Jødal, A.-S. S., and Larsen, K.L. 2021. Investigation of the relationships between the alveograph parameters. *Sci. Rep.*, 11:5349.
39. Brandner, S., Becker, T., and Jekle, M. 2019. Classification of starch-gluten networks into a viscoelastic liquid or solid, based on rheological aspects—A review. *Int. J. Biol. Macromol.*, 136:1018.
40. Joshi, Y.M., and Denn, M.M. 2004. Failure and recovery of entangled polymer melts in elongational flow. *Rheol. Rev.*:1–17.
41. Grenier, D., Rondeau-Mouro, C., Dedey, K.B., Morel, M.-H., and Lucas, T. 2021. Gas cell opening in bread dough during baking. *Trends Food Sci. Tech.*, 109:482.
42. Chen, Y.R., Seabourn, B.W., and Herald, T.J. 2015. Modified dough preparation for Alveograph analysis with limited flour sample size. *Cereal Chem.*, 92:565.

Index